ARTS, ECOLOGIES, TRANSITIONS

Arts, Ecologies, Transitions provides in-depth insights into how aesthetic relations and current artistic practices are fundamentally ecological and intrinsically connected to the world. As art is created in a given historic temporality, it presents specific modalities of productive and sensory relations to the world.

With contributions from 49 researchers, this book tracks evolutions in the arts that demonstrate an awareness of the environmental, economic, social, and political crises. It proposes interdisciplinary approaches to art that clarify the multiple relationships between art and ecology through an exploration of key concepts such as collapsonauts, degrowth, place, recycling, and walking art. All the artistic fields are addressed from the visual arts, theatre, dance, music and sound art, cinema, and photography – including those that are rarely represented in research such as digital creation or graphic design – to showcase the diversity of artistic practices in transition.

Through original research this book presents ideas in an accessible format and will be of interest to students and researchers in the fields of environmental studies, ecology, geography, cultural studies, architecture, performance studies, visual arts, cinema, music, and literature studies.

Roberto Barbanti is Emeritus Professor in the Department of Visual Arts at Université Paris 8 and a member of the Arts des images et art contemporain research unit. He co-founded and co-edited the journal *Sonorités* (2006–2017) and is an advisory board member for the publisher Eterotopia France. His research areas cover ecosophy, sound ecology, and contemporary art. His publications include: *Les limites du vivant* (co-edited with Lorraine Verner, 2016), *Dall'immaginario all'acustinario. Prolegomeni a un'ecosofia sonora* (2020), and *Les sonorités du monde. De l'écologie sonore à l'écosophie sonore* (2023).

Isabelle Ginot is Professor of Dance Studies at Université Paris 8 and co-founder of the association Association d'individus en mouvements engagés. Her two main areas of research intersect with issues of vulnerability and difference in dance. The first addresses dance performance analysis and criticism, and focuses on artists with disabilities who perform on stage. The second analyses practices, especially practices (workshops, performances, participatory art) with "non-dancers" who have disabilities, are ageing, or are affected by social discrimination and exclusion. From 2010 to 2019, she chaired the research group Soma&Po, developing research and practices on the political and social uses of somatic practices. Currently, she runs a practice-based seminar entitled "Mouvements engagés" (Engaged moves), a peer-led workshop that shares dance practices in French care institutions. Lastly, she investigates alternative formats for academic research, involving the participation of artists, activists, and non-scholar actors.

Makis Solomos was born in Greece and lives in France. He is Professor of Musicology at Université Paris 8 and Director of the research unit MUSIDANSE. He has published many books and articles about new music. His main areas of research are the focus on sound, the notion of musical space, new musical technics and technologies, the mutations of listening, and the ecology of sound. His book *From Music to Sound: The Emergence of Sound in 20th- and 21st-Century Music* (Routledge, 2019) deals with an important change in today's music. His latest book *Towards an Ecology of Sound: Environmental, Mental and Social Ecologies in Music, Sound Art and Artivisms* (Routledge, 2023) addresses an expanded notion of ecology, mixing environmental issues and socio-political questions. He is also one of the main Xenakis specialists, to whom he has devoted many publications. For Xenakis's centenary (2022), he co-organized the "Xenakis22: Centenary International Symposium" and he is the editor of *Révolutions Xenakis* (Éditions de l'Œil – Philharmonie de Paris, 2022).

Cécile Sorin is currently Professor in the Department of Cinema at Université Paris 8. After having published books on practices of parody and pastiche in cinema (*Pratiques de la parodie et du pastiche au cinéma*, 2010) and Pasolinian pastiche (*Pasolini, pastiche et mélange*, 2017), she is currently examining processes of subjectivation in Pasolinian cinema and contemporary French cinema. Her recent work reassesses Pasolini's work through the prism of ecopoetics. In addition, she is co-editor of the "Esthétiques hors cadre" series from Presses Universitaires de Vincennes.

ARTS, ECOLOGIES, TRANSITIONS

Constructing a Common Vocabulary

Edited by Roberto Barbanti, Isabelle Ginot,
Makis Solomos, and Cécile Sorin

LONDON AND NEW YORK

Designed cover image: Yann Aucompte

First published 2024
by Routledge
4 Park Square, Milton Park, Abingdon, Oxon OX14 4RN

and by Routledge
605 Third Avenue, New York, NY 10158

Routledge is an imprint of the Taylor & Francis Group, an informa business

British Library Cataloguing-in-Publication Data
A catalogue record for this book is available from the British Library

Library of Congress Cataloging-in-Publication Data
Names: Barbanti, Roberto, editor. | Ginot, Isabelle, editor. | Solomos, Makis, editor. | Sorin, Cécile, editor.
Title: Arts, ecologies, transitions / edited by Roberto Barbanti, Isabelle Ginot, Makis Solomos and Cécile Sorin.
Description: Abingdon, Oxon : Routledge, 2024. | Includes bibliographical references and index. | Contents: Acoustic ecology / Kostas Paparrigopoulos -- Learning and experience / Anastasia Chernigina and Antoine Freychet -- Walking art / Antoine Freychet and Anastasia Chernigina.
Identifiers: LCCN 2023042380 (print) | LCCN 2023042381 (ebook) | ISBN 9781032595153 (hardback) | ISBN 9781032596143 (paperback) | ISBN 9781003455523 (ebook)
Subjects: LCSH: Environment (Aesthetics) | Ecology in art.
Classification: LCC BH301.E58 A785 2024 (print) | LCC BH301.E58 (ebook) | DDC 701/.17--dc23/eng/20231205
LC record available at https://lccn.loc.gov/2023042380
LC ebook record available at https://lccn.loc.gov/2023042381

ISBN: 978-1-032-59515-3 (hbk)
ISBN: 978-1-032-59614-3 (pbk)
ISBN: 978-1-003-45552-3 (ebk)

DOI: 10.4324/9781003455523

Typeset in Times New Roman
by SPi Technologies India Pvt Ltd (Straive)

Translation: Stacie Allan

Script reader: Johnny McFadyen

This research has been supported by EUR ArTeC and financed by the French National Agency for Research (ANR) through the PIA ANR-17-EURE-0008 and by the research centres MUSIDANSE, ESTCA, and AIAC of University Paris 8.

We dedicate this book to migratory birds.

CONTENTS

CONTRIBUTORS

Pavlos Antoniadis is Associate Professor in Communications and Music Technology at the University of Ioannina. He received his PhD from the Université de Strasbourg and Institut de recherche et coordination acoustique/ musique (IRCAM). He collaborates with the Sound Music Movement Interaction team at IRCAM and is a member of the ACCRA research centre at the Université de Strasbourg. He previously held postdoctoral positions at the EUR-ArTeC research centre at Université Paris 8 and the Technische Universität Berlin (Audio Communication) as a Humboldt Stiftung scholar. As a pianist, musicologist, and technologist, Antoniadis's research explores complexity, embodied cognition, multimodality, interactive technologies, motion capture, computational musicology, epistemology of performance, AI, and biopolitics in human and machine learning, with a focus on the repertoire of the twentieth and twenty-first centuries. He has recently completed two monographs, published by Wolke Verlag, and several articles. Amongst his other activities, he has developed the software GesTCom with Fred Bevilacqua, and presented Iannis Xenakis's complete works for the piano through augmented reality in collaboration with Makis Solomos and Jean-François Jégo.

Rémi Astruc is Professor of Comparative Literature in the Department of Literature at the CY Université Cergy Paris, and a member of the research unit Héritages. With RKI Press, he has published two books *Le design de la communauté* (2023) and *Nous? L'aspiration à la communauté et les arts*, with a preface written by Jean-Luc Nancy (2016). He also edited the volume *COMMUNITAS, Les mots du commun* (2020) and founded the Communauté des Chercheurs sur la Communauté (CCC) network whose blog he writes: https://communaute deschercheurssurlacommunaute.wordpress.com/

Yann Aucompte holds a PhD in aesthetics and the science of art from Université Paris 8. He teaches graphic design as part of a national diploma in art and design, specialising in book design and publishing, and storytelling and communicating science through graphic design and illustration. His research interests lie in the relationship between graphic design and politics, philosophy, ecology, and hospitality towards asylum seekers, as well as ecosophy and theories of complexity. He is a co-editor of *La Querelle de la Déconstruction: Un débat philosophique dans le Design, Graphique? Une anthologie transatlantique* (T&P Publishing, 2023) with Stéphane Darricau. His publications have appeared in the journals *Sciences du design*, *Appareil* (with Stéphane Darricau), and *Cultures Visuelles*, and the volumes *Entre les lignes. Anthropologie, littérature, arts et espace* (2021) and *Le discours critique en art & en design. Pratiques et enjeux contemporains* (2022).

Marie Bardet is originally from France and is an academic in the Interdisciplinary School of Advanced Social Studies at the Universidad Nacional de General San Martín, Buenos Aires where she leads the master's programme in contemporary artistic practices in the School of Art and Heritage. Her projects and research shift the boundaries between theory and practice and are enriched as much by improvisation in dance and somatic practices as by contemporary philosophy and queer/cuir feminist thought practices. She is interested in issues that, through movement and sensory materials, make up common spaces where situated thinking, artistic multiplicity, and political concerns are at once exercised. Bardet supports processes of research and creation and regularly gives seminars at different universities and museums (Museo Universitario Arte Contemporaneo Mexico, Maestria interdisciplinary en Teatro y Artes vivas Bogotá, Museo Reina Sofía). Her publications include *Improvisación en danza. Traducciones y Distorsiones* (2022), *Perder la cara* (2021), *Una paradoja moviente: Loïe Fuller* (2021), *Hacer Mundos con Gestos* (2019), *Pensar con Mover* (2012), and *Penser et Mouvoir* (2011).

Eliane Beaufils is Associate Professor in Theatre Studies at Université Paris 8, a member of the research unit "Scènes du monde", and Assistant Director of the doctoral college Ecole doctorale Esthétique, sciences et technologies des arts (EDESTA). Her research focuses on contemporary drama and performance in Europe, participatory art, spectatorial autopoiesis, and the themes of criticism and resonance. Following a project on new theatre criticism (published as *Toucher par la pensée*, Hermann, 2021), she began a multi-year research project on theatre and the climatic future in 2018, and later a second project on possible world stages (2022–2024). Amongst other works, she is the author of *Violences sur les scènes allemandes* (2010), *Quand la scène fait appel … Le théâtre contemporain et le poétique* (2014), *Being-With in Contemporary Performing Arts* (with Eva Holling, 2018), and *Scènes en partage* (with Alix de Morant, 2018).

Aline Bergé is Associate Professor at the Université Sorbonne Nouvelle, and a member of the research unit Théorie et histoire des arts et des littératures de la modernité (THALIM) and the transition training collective Formation à la Transition écologique et sociale dans l'Enseignement Supérieur FORTES). Author of the *Manuel de la Grande Transition* (Les Liens qui Libèrent, 2020), Bergé co-directs a research strand on environmental humanities and transitions at the Sorbonne Nouvelle. Her research addresses the ecopoetics of French and francophone literatures at the intersection of arts and knowledge about living beings. Her other publications include: *Philippe Jaccottet, trajectoires et constellations – Lieux, livres, paysages* (2004); co-editing *Paysage & Modernité(s)* (2007) with Michel Collot and *Paysages européens et Mondialisation* (2012) with Michel Collot and Jean Mottet; and writing the section "Littérature française XXe (France et francophonie)" with Bernard Alazet, in the *Dictionnaire universel des Créatrices* (2013).

Augustin Berque is a French geographer and retired director of studies at the École des hautes études en sciences sociales, Paris. A member of the Academia europaea and honorary member of the European Association for Japanese Studies, he was the first Western scholar to receive the Grand Prize of Fukuoka for Asian cultures (福岡アジア文化賞大賞) in 2009. For his contributions to mesology following on from Uexküll's *Umweltlehre* and Watsuji's *fûdogaku*, he was also awarded the International Cosmos prize in 2018. His publications include *Thinking through Landscape*, translated by A.-M. Feenberg (Routledge, 2015).

Nathalie Blanc is Director of Research at the Centre National de la Recherche Scientifique, Director of the Centre des Politiques de la Terre, and former director of the research unit Laboratoire dynamiques sociales et recomposition des espaces (LADYSS) (2014–2019). A pioneer of ecocriticism in France, she has published and coordinated research programmes on areas such as nature in the city, environmental aesthetics, and environmental activism. A founding member of the online portal Humanités Environnementales, she was also the French delegate for the European research network Programme de cooperation européenne en science et technologie (COST) "Investigating cultural sustainability" (2011–2015) and delegate for the COST programme on new materialisms "How Matter Matters" (2016–2019). More recently, her work has focused on the theme of environmental activism, with a territorial project on the contribution of mobilisations to the territories, a cultural project on the values inherent to these mobilisations, and a political project on territorial politicisations. The place of creativity and art is examined within these dynamics of social innovation. Her recent publications include *Art, Farming and Food for the Future: Transforming Agriculture* (Routledge, 2023) and *Réparer la Terre par le bas. Manifeste pour un environnementalisme ordinaire* (2022).

Clara Breteau is Associate Professor in Arts and Ecology at Université Paris 8. With a PhD in geography and aesthetics, she holds degrees from the universities of Cambridge, Leeds, and the Sorbonne. She has published on political and poetic ecology in *Les Vies autonomes, une enquête poétique* (2022), with a preface by Camille de Toledo. Approaching autonomous ecological lifestyles as "houses that are eaten", she shows that they revive, in the folds of their metabolism, the existential, poetic, and political strengths of our vernacular worlds. Breateau is also an author of poems, short stories, articles, and a documentary *Les Maisons légères*, produced for Radio France and broadcast on France Culture in 2015.

Giusy Checola is a PhD candidate at Université Paris 8. In parallel, she collaborates with the Institute for Public Art (UK) as a researcher and regional representative for Europe. She is a member of the transdisciplinary Space Ecologies Art and Design network and is on the advisory board of Fonds Roberto Cimetta. She has worked on the development and delivery of multi-year projects on artistic renovation of cultural, architectural, and environmental heritage, supported by EU programmes such as Creative Europe, the European Regional Development Fund, and European Capital of Culture 2019. She has authored and edited books published by Routledge, Mimesis Éditions, and Postmedia Books.

Anastasia Chernigina is a PhD candidate at Université Paris 8. A linguist and former foreign languages teacher from Moscow, she pursued postgraduate studies at the Université de Franche-Comté and a master's degree in Education Sciences at Paris 8 where she began her action research project on sound walk installations and sound creations. She expanded and developed the pedagogical aspect of her installations alongside the musicologists Antoine Freychet and Raphael Bruni. Her existential research on the link between humans and their environment became her doctoral project in Education Sciences, co-supervised in Musicology from 2016 onwards. Since then, Chernigina has worked as a teacher and facilitator of these practices for teenagers in different organisations and places.

Philippe Chiambaretta, after training in science and economics at l'École des Ponts et Chaussées in Paris and MIT in Boston, worked in strategic consultancy for Booz Allen Hamilton and managed international activity for Ricardo Bofill Taller de Arquitectura in Paris. A graduate of the École Nationale Supérieure d'Architecture Paris-Belleville (2000), he created PCA, a research and architectural agency characterised by the synergy between thought and action. Resolutely turned towards innovation, the city of tomorrow, and new usages, the agency PCA-STREAM is a creative and transdisciplinary ecosystem that today brings together a pluridisciplinary team of 90 collaborators with different profiles (architects, urbanists, designers, engineers, researchers, editors) capable of understanding and responding to increasingly complex challenges in the contemporary world.

Yves Citton is Professor of Literature and Media at Université Paris 8. Until 2021, he was Executive Director of the university research school ArTeC. He co-edits the journal *Multitudes* and is the author of *Altermodernités des Lumières* (2022), *Faire avec. Conflits, coalitions, contagions* (2021), *Générations Collapsonautes. Naviguer en temps d'effondrements* (with Jacopo Rasmi, 2020), *Contre-courants politiques* (2018), *Médiarchie* (2017), *Pour une écologie de l'attention* (2014), and *Zazirocratie* (2011). His articles are available on www.yvescitton.net.

Joanne Clavel is a researcher at the Centre national de la recherche scientifique (CNRS) within the Laboratoire dynamiques sociales et recomposition des espaces (LADYSS) research group at the Université Paris Cité and is an associate of the Department of Danse at Université Paris 8. She approaches ecological humanities from knowledge about the body. Trained as an ecologist, she studied the impact of global changes on biodiversity and biotic homogenisation at the Muséum national d'Histoire naturelle for almost ten years. She then trained in environmental humanities and dance research to examine somatic questions and the politics of the disappearance of living beings and ecosystemic transformations. She has carried out studies with different actors including farmers from the world of art and the world of agriculture. With the group Soma&Po, she coordinated the "Écosomatiques, Penser l'écologie depuis le geste" conference in 2014 at the Centre National de la Danse and the subsequent book (2019). She continues to contribute to research on theories of movement through writings on environmental aesthetics, studies on the links between art and ecologies, and a practice of research creation in close collaboration with choreographers.

Gilles Clément is a French gardener, landscaper, botanist, entomologist, biologist, and writer. After training as a horticultural engineer (1967) and landscaper (1969), he began teaching at the École nationale supérieure du paysage in Versailles in 1979, whilst also working as a designer. His design of the Parc André-Citroën in Paris (1992), the spectacular exhibition *Le Jardin planétaire* (1999) which he commissioned at the Grande Halle de la Villette, and his theoretical and literary writings have brought him to the attention of the wider public. He is the author of numerous books including *Notre-Dame-des-Plantes* (2021), *Jardins, paysage et génie naturel* (2012), *Le Salon des berces* (2009), *Le Jardin en mouvement: de la vallée au jardin planétaire* (2001), *Le Jardin planétaire* (1999), *La Dernière Pierre* (1999), *Les Libres Jardins* (1997), and *Thomas et le voyageur: esquisse du jardin planétaire* (1997).

Nathalie Coutelet is Professor in the Department of Theatre at Université Paris 8. Her research interests lie in reduced forms of live performance and their political dimension. She is the author of *Un Théâtre à côté: la Grimace* (2020) and *Etranges artistes sur la scène des Folies-Bergère* (2015), and co-edited *L'Altérité en spectacle* (2015) with Isabelle Moindrot. She is currently working on a project about historical and contemporary "theatres of nature", in collaboration with the Réseau Européen des Théâtres de Verdure and the Université de Lausanne.

Agostino Di Scipio is a composer, sound artist, and researcher. He is Professor of Electroacoustic Music Composition in L'Aquila, having previously held the same position at the conservatoire in Naples. He studied composition and electronic music at the conservatoire in L'Aquila (Italy) and spent time at the Centre for Computer Sound at the Università degli Studi di Padova before receiving his PhD from Université Paris 8. His musical production work includes pieces for instruments with electroacoustic and audio digital live performances, sound installations, and purely instrumental pieces. He works with non-conventional practices to generate and transmit sound, often centred on the relationship between humans, machines, and the environment. He has been a visiting lecturer for different institutions (Deutscher Akademischer Austauschdienst German Academic Exchange Service DAAD in Berlin, Technische Universität de Berlin) and is the author of studies on the histories of technologies of sound and music and their cognitive and cultural implications. His most recent publication is *Circuiti del tempo. Un percorso storico-critico nella creatività elettroacustica e informática* (2021).

Frédérick Duhautpas is Associate Professor in the Department of Music at Université Paris 8 where he teaches classes on feminist musicology, musicology and social sciences, and history and music aesthetics at the turn of the twentieth century. He is affiliated with the research unit MUSIDANSE at Paris 8. His research addresses the questions of expressivity, signification in music, and relationships between music and society, notably questions of gender and sound ecology. He has also worked on the music of Hildegard Westerkamp, Claude Debussy, Nadia Boulanger, Lili Boulanger, Iannis Xenakis, and Luciano Berio, amongst others.

Ludovic Duhem is an artist and philosophy specialist. He is currently research coordinator at the Ecole supérieure d'art et de design de Valenciennes (ÉSAD) de Valenciennes where he teaches philosophy of art and design, as well as at the universities of Lille and Valenciennes, the Ecole nationale supérieure des arts visuels de La Cambre (ENSAV-La Cambre), and the Ecole nationale supérieure de creation industrielle (ENSCI-Les Ateliers). His research interests lie in the relationship between aesthetics, technology, and politics within contemporary ecological issues. He recently published *Crash metropolis. Design écosocial et critique de la métropolisation des territoires* (2022), *Écologie et technologie. Redéfinir le progrès après Simondon* (with Jean-Hugues Barthélémy, 2022), and *Les écologies du numérique* (2022).

Matthieu Duperrex is Associate Professor in Humanities at the École nationale supérieure d'architecture de Marseille and part of the Inama research centre. Artist, theorist, and artistic director of the collective *Urbain, trop urbain* (www. urbain-trop-urbain.fr), his research is based on fieldwork in anthropized environments and overlaps with literature, the humanities, and visual arts. His publications include *Voyages en sol incertain, enquête dans les deltas du Rhône et du Mississippi* (2019), *Semer le trouble. Soulèvements, subversions, refuges* (2020), *Fos - Étang de Berre. Un littoral au cœur des enjeux environnementaux* (2021), and *La rivière et le bulldozer* (2022).

Claire Fagnart worked as Associate Professor (1995–2023) in the Department of Visual Arts at Université Paris 8 and was a member of the Esthétique, Pratique et histoire des Arts (EPHA) group within the AIAC research unit. She led a research strand about languages and discourses on art until 2022 and carried out a project on art criticism (2012–2015). One of her research areas concerns the question of distance envisaged in a philosophical perspective, viewpoints of creation and artist reception, and sensory viewpoints. She also works on the beliefs and ideologies implicated in the relationships between art works and written discourse, and in the relationships between art works and documents. Her publications include *La Critique d'art* (2017), *L'Art au XXe siècle (et) l'utopie* (2000) with Roberto Barbanti, and *La Critique d'art en question* (2014).

Antoine Freychet studied saxophone and musicology in Angers. At Université Paris 8, he obtained master's degrees in music and philosophy before writing the doctoral thesis which forms the basis of his book *Démarches artistiques et préoccupations écologiques. L'écoute dans l'écologie sonore* (2022). Passionate about sound ecology, he participated in different projects (collective projects on arts, ecologies, transitions, and the usages of sound) and edited an Argentinian collective work and volume of the journal *Filigrane. Musique, esthétique, science, société sur le field recording*. Attentive to nature as well as human environments, he practised the art of sound walks, created sound installations, and played saxophone as part of the group Bilboquet. He passed away in August 2022.

Anne-Laure George-Molland is Associate Professor in the Department of Cinema at the Université Paul-Valéry Montpellier 3 and a member of the interdisciplinary research in creation strand within the Représenter, inventer la réalité, du romantisme au XXIè siècle (RIRRA21) research group. Having followed an art–science pathway and specialising in computer graphics, she contributed to research that addresses the process of creation in animation studios and how they shape, develop, and are remodelled in contact with works and contexts of production. Conscious of environmental issues, she took a new direction in her research and teaching by starting from an ecosophic reassessment of digital creation to examine new practices, forms, and artistic works. Since 2021, she has participated in the "Arts, Écologies, Transitions" project at Université Paris 8, as well as the development and implementation of a cross-disciplinary option in ecological humanities at Montpellier 3.

Alice Gervais-Ragu is working on a thesis entitled "Ecologie des nouveaux imaginaires chorégraphiques", supervised by Isabelle Ginot at Université Paris 8, which examines relational dramaturgy and the modes of existence of plays in a generalised context of crisis. She teaches analysis of works and dramaturgy (universities of Paris 8, Paris 1 Panthéon-Sorbonne, Cergy) and speaks at different conferences and training programmes (Université Nice-Sophia-Antipolis; the practice-based master's programme at CCN de Montpellier). Her publications include contributions to *Art et mathématiques* (2020), *Danse contemporaine et littérature* (2017), and *Danse et education* (2016). She regularly writes for the MaCulture website, *Journal de l'ADC*, and *Repères Cahiers de danse*. She has also published two poetry collections *Reprendre trois fois de tout* (2017) and *La dernière forêt* (2023), and is writing a novel for which she received a bursary from the Centre National du Livre.

Gala Hernández López is an artist and researcher. She is a PhD candidate at the EDESTA doctoral college and Esthétique, sciences et technologies du cinema et de l'audiovisuel (ESTCA) research centre at Université Paris 8 where she is developing a research-creation project on screenshots as media in the post-internet age and where she taught for three years. She is currently a teaching fellow at the Université Gustave Eiffel and was a visiting researcher at the Filmuniversität Babelsberg Konrad Wolf thanks to a doctoral research scholarship from DAAD. She co-founded the research and art creative Après les Réseaux Sociaux and was selected for Berlinale Talents 2023. Her work brings together interdisciplinary research with the production of essay films on new modes of subjectivation produced specifically by digital capitalism. Her work has been shown at international festivals such as DOK Leipzig, Cinéma du Réel, Clermont-Ferrand, and IndieLisboa.

Jean-François Jégo is Associate Professor in the Art and Imaging Technology department of the Faculty of Arts at Université Paris 8. He is also a digital artist and researcher at the Images numériques et réalité virtuelle (INReV)

Virtual Reality Laboratory. Following a transdisciplinary master's degree in contemporary arts, Jégo obtained a PhD in computer science and virtual reality at the robotics centre of Mines ParisTech. Over the last 15 years, he has created art installations and interactive theatre performances that bring together virtual and augmented reality. His current research in digital art addresses the ecosophy and aesthetics of interaction for both the artist and spectator. He has created award-winning digital art installations and co-scripted digital theatre performances broadcast at different international events. He co-founded the Virtual Reality Art Collective. His articles, resources, and portfolio can be found on: www.jfcad.com

Susana Jiménez Carmona is a sound artist and researcher with a PhD in Humanities and Cultural Studies. A philosophy graduate and musician, she teachers on the Sound Art master's programme at the Universitat de Barcelona. Her work is based on collaborative practices, interspecific listening, and the relationship between art, science, and politics. She is a member of different collectives (*cuidadoras de sonidos, el paseo de Jane*) and works with artists and activists from Spain and abroad. She has also written on composers and artists, such as Luigi Nono, Jana Winderen, Félix Blume, Janet Cardiff, Ultra-red, Morton Feldman, and Hildergard Westerkamp.

Isabelle Launay is Professor in the Department of Dance at Université Paris 8 and a member of the MUSIDANSE research unit. Her most recent publications address the memory of works in dance in France: *Poétiques et politiques des répertoires, Les danses d'après 1* (CND, 2017) and *Cultures de l'oubli et citation, Les danses d'après 2* (CND, 2018). For many years, she has contributed to the training of choreographic artists (Centre national de danse contemporaine, CNDC in Angers, the practice-based master's programme at the CCN de Montpellier) and collaborated with artists on various projects within the field of contemporary dance including Mathilde Monnier, Loïc Touzé, Emmanuelle Huyhn, Cécile Proust, Latifa Laabissi, Bintou Dembélé, and Lia Rodrigues. Her other publications include: *À la recherche d'une danse moderne, Rudolf Laban et Mary Wigman* (1996); *Entretenir, à propos d'une danse contemporaine* (2002) with Boris Charmatz; *Les Carnets Bagouet, la passe d'une œuvre* (2008); *Mémoires et histoire en danse* (2011) with Sylviane Pagès; *Histoires de gestes* (2012) with Mari e Glon; and *La passion des possibles, Lia Rodrigues, 30 ans de compagnie* (2021) with Silvia Soter.

Eric Lecerf is Associate Professor of Philosophy at Université Paris 8. As a specialist in theories of emancipation, he has published several books on the subject of labour and numerous articles on authors within the libertarian tradition such as Pierre-Joseph Proudhon, Georges Sorel, and Simone Weil, as well as on the philosophers Henri Bergson and Günther Anders.

Alberto Magnaghi is President of the multidisciplinary association, the Territorialist Society. As founder of the Italian Territorialist School, he has coordinated national research projects and experimental laboratories for the Ministry of Universities and the National Research Council on the topics of self-sustainable local development, representing the identity of the territory, environment and landscape, the territory project, and urban bioregions. He has acted as a scientific coordinator and designer for urban, territorial, and landscape projects and plans for a strategic and integrated natural environment. For different municipalities, he has coordinated and designed participatory projects and plans for the social production of the territory and landscape. His most recent publications include *Il principio territorial* (2020) and, as editor, *La regola e il progetto: un approccio bioregionalista alla pianificazione territoriale* (2014).

Damien Marguet is Associate Professor and Co-director of the Department of Cinema at Université Paris 8. He is a member of the ESTCA research centre, a film maker, and a scheduler. His doctoral thesis explored the poetics of translation in the works of Pier Paolo Pasolini, Danièle Huillet and Jean-Marie Straub, and Béla Tarr. A member of the research group "Théâtres de la mémoire", his work addresses: eastern European cinema, in particular Hungarian cinema; the relationship between cinema, literature, language, and translation; cinema and ecology; and the questions of duration, action, and experience in the field of moving images. He is the co-author of *Béla Tarr. Les Métamorphoses d'un visionnaire* (2022) and he co-edited *Serguei Loznits cinéma à l'épreuve du monde* (2022). Marguet is also a director of experimental films, a member of Collectif cinemacinéma (CJC) and Labo K, and serves on the editorial board of the journal *Passés Futurs*.

Baptiste Morizot is a philosophy specialist and prize-winning writer. His philosophical research is centred on the place of humans amongst living beings. His book *Les diplomates. Cohabiter avec les loups sur une autre carte du vivant* received the Fondation de l'écologie politique's book prize in 2016 and the Prix Littéraire François Sommer in 2017. *Sur la piste animale* (2018), which also addressed animal tracking through different stories, received the Académie française's Prix Jacques-Lacroix in 2019. *Esthétique de la rencontre. L'énigme de l'art contemporain* (2018), written in collaboration with Estelle Zhong Mengual, received the Rencontres philosophiques d'Uriage prize in 2019. In *Manières d'être vivant: Enquêtes sur la vie à travers nous* (2020), he broadened a series of philosophical investigations based on the practice of tracking. His latest book is entitled *L'inexploré* (2022).

Kostas Paparrigopoulos is a musicologist who researches the music of milieu from the twentieth century to the present, in particular the music of Iannis

Xenakis and John Cage. He is interested in the links between music and social and ecological questions, as well as theories that emphasise an ecological conceptualisation of living beings such as ecosophy (Félix Guattari), degrowth (Serge Latouche), and transition (Rob Hopkins). He is part of the AIMEE group and the project "Arts, Écologies, Transitions", initiatives that target current ecological transitions, ethics-aesthetics of the environment, the production of subjectivity, and the concept of milieu. He is currently Vice President of the Hellenic Society for Acoustic Ecology, an organisation affiliated with the World Forum for Acoustic Ecology.

Carmen Pardo Salgado is Professor at the Universitat de Girona and teaches on the Sound Art master's programme at the Universitat de Barcelona. Her research interests lie in questions of listening and voice, ecosophy and sound ecology, as well as contemporary music and the aestheticisation of the social. A specialist in John Cage, she has written several books on him, including *John Cage, Escritos al oído* (1999) and *La escucha oblicua: una invitación a John Cage* (2014, French version *Approche de John Cage. L'écoute oblique*, 2007, winner of the Académie Charles Cros's Coup de cœur award 2008). She is also author of: *Música y Pensamiento, apuntes de un encuentro* (2019); *En el silencio de la cultura* (2016; French version: *Dans le silence de la culture*, 2018); *Robert Wilson* (with Miguel Morey, 2003); and *Las TIC: una reflexión filosófica* (2009).

Soko Phay is Professor of History and Theory of Contemporary Art at Université Paris 8 and at the New College of Political Studies, Université Paris Lumières. Since 2019, she has directed the AIAC research centre. Her research addresses the aesthetics of the mirror (*Les vertiges du miroir dans l'art contemporain*, Les presses du réel, 2016) and art in the face of the extreme in its relationship to memory and history. She co-edited *Cambodge, cartographie de la mémoire* (2017) with Patrick Nardin and Suppya Nut; *Archives au présent* (2017) with Patrick Nardin, Catherine Perret, and Anna Seiderer; *Les génocides oubliés?* (2020) and *Rwanda, l'atelier de la mémoire. Des archives à la création* (2022) with Pierre Bayard; and *Le paysage après coup* (2022) with Patrick Nardin.

Michel Poivert is Professor of Art History at Université Paris 1 Panthéon-Sorbonne where he holds a chair in the history of photography. In 2018, he established a project at the Collège international de philosophie on the transmission and experimentation of photographic skills. He is also a critic and exhibition curator. Amongst other works, he is the author of *La photographie contemporaine* (2018), *50 ans de photographie française de 1970 à nos jours* (2019), and *Contreculture dans la photographie contemporaine* (2022). He organised the exhibitions "L'Événement, les images comme acteur de l'histoire" (2007), "Gilles Caron Paris 1968" (Hôtel de Ville, Paris, 2018), and

"La photographie française: une métamorphose 1968–1989" (Pavillon populaire de Montpellier, 2022).

Jacopo Rasmi is Associate Professor of Visual Arts and Italian Studies at the Université Jean Monnet – Saint Etienne. He is the author of *Le hors champ est dedans. Michelangelo Frammartino, écologie et cinéma* (2021), *Générations Collapsonautes. Naviguer par temps d'effondrement* (2020) with Yves Citton, and *Studium* (2019) with François Deck. From time to time, he organises screenings of documentary cinema. Even in these screenings or his single-authored texts, collective intelligence, such as that found in writings in *Multitudes* or *La revue documentaires*, support and enrich his work.

Fabrice Rochelandet is Professor of Communication Sciences at Sorbonne Nouvelle University and a specialist in the cultural and digital economy. He has published works in the fields of economics and communication and information systems on cultural industries and the media (cinema, online press), in particular processes of platformisation, digital business models, and their regulation (copyright, personal data, cultural diversity). He has coordinated research projects funded by the National Agency for Research addressing collaborative networks and the illegal sharing of cultural commons, digital distribution models for audio-visual content, and more recently creative territories and the audio-visual industry.

Matthieu Saladin is an artist and Associate Professor of Visual Arts at Université Paris 8 where he is a member of the TEAMeD group within the AIAC research unit. His practices take a conceptual approach to art, reflecting on the production of spaces, the history of creative forms and processes, and the relationship between art and society from an economic and political viewpoint. They also take the form of protocols, installations and performances, publications (books, records), video, and software creations. His theoretical research mainly addresses sound art and experimental music. He co-edits the series "Ohcetecho", serves on the editorial board for the journals *Volume!* and *Revue et Corrigée*, and was editor of the research journal *TACET*. His work is represented by the gallery Salle Principale.

Julie Sermon is Professor of History and Aesthetics of Contemporary Theatre at Université Lyon 2 and Director of the Passages Arts & Littératures (XX–XXI) research centre, within which she facilitates the environmental humanities strand. She is author of several works on the renewal of languages, forms, and textual and theatre practices, in particular exploring the phenomena of decentring (theory, aesthetics, action) that they imply. Since 2017, she has mainly focused her teaching and research activities on binding relations, in both meanings of the word, between theatrical arts and ecology. Her work on the subject is published in *Morts ou vifs. Contribution à une écologie sensible, théorique et pratique des arts vivants* (2021).

Hélène Singer holds a PhD in art and the science of art from the Université Paris 1 Panthéon-Sorbonne. As an artist-researcher, she primarily explores voice, animality, and the body in action in contemporary art. Her monograph *Expressions du corps interne. La voix, la performance et le chant plastique*, based on her doctoral thesis, was published in 2011, and she edited a special issue of *Revue Ligeia, Dossiers sur l'art* on art and animality (2016) in partnership with Paris Musée. Having held various teaching positions (universities of Lille 3, Paris 1, and Paris 8), she currently teaches photographic arts at the École des Beaux-Arts de Versailles. She has exhibited her pluridisciplinary works in art centres and museums and performs as a singer in collaboration with composers and musicians.

Thierry Tremblay, with a PhD from Université Paris VII, was Senior Lecturer at the University of Malta. He previously taught Literature, Philosophy, and the History of Ideas at Charles University and the Anglo-American University in Prague, as well as the University of Cyprus. With the publisher Hermann, he published *Anagogiques, De la transgression aux sommets* (2022) and *Anamnèses: Essai sur l'œuvre de Pierre Klossowski* (2012). He is also the author of *Frontières du sujet: Une esthétique du déclin* (2015) and a collection of poems *Comme* (2023). He passed away at the end of 2022.

Lorraine Verner is Research Associate at Université Paris 8 within the AIAC research unit and is a member of the project team for Arts, écologies, transitions. She received her PhD from the Centre d'Histoire et de Théorie des Arts at the École des hautes études en sciences sociales, was a post-doctoral researcher at the Institute of Art History at the Freie Universität Berlin, and held teaching positions in various higher education institutions. Since 2011, she has taught at the École des Beaux-Arts de Versailles. Her research, which addresses art and ecosophy with a particular interest in the sciences of nature, has been published in several books, including *Les limites du vivant. À la lisière de l'art, de la philosophie et des sciences de la nature* (2016) which she co-edited. She was part of the residency "Au pays de Herr Joseph Beuys. Contexte(s)" at ArToll Kunstlabor in Germany from 2014 to 2020.

Tiziana Villani is a philosophy specialist, who teaches the phenomenology of contemporary art at NABA, Milan and Sapienza Università di Roma. She is an editor at Eterotopia France and for the journal *millepiani* in Italy. Her publications include: *Athena Cyborg. Per una geografia dell'espressione: corpo,* *territorio, metropoli* (1995); *Gilles Deleuze. Un filosofo dalla parte del fuoco* (1998); *Il tempo della trasformazione* (2006); *Ecologia Politica,* (2013); *Psychogéographies urbaines. Corps, territoires et technologies* (2014); *Corps mutants. Technologies de la sélection de l'humain et du vivant*, Paris (2019); "Telós. Costruire progetti, alleanze, relazioni", in *Prendiamo corpo* (2021); and *Le crepe del presente* (2023).

Estelle Zhong Mengual holds the chair "Habiter le paysage – l'art à la rencontre du vivant" at the École des Beaux-Arts in Paris. She is a historian of art and received her PhD from Sciences Po Paris where she teaches on the master's programme "Expérimentation en Art et Politique", created by Bruno Latour. Her current research addresses the relationship that art, past and present, maintains with the living world. In particular, she is developing an environmental history of art, which proposes a new regime of attention to the representation of living beings in art using tools from innovations in the environmental humanities and natural sciences. She is the author of *Apprendre à voir. Le point de vue du vivant* (2021), which won the EcoloObs prize for the best essay on environmental thought published in 2021, and *Peindre au corps à corps. Les fleurs et Georgia O'Keeffe* (2022).

PREFACE: A COLLECTIVE – ARTS, PRACTICES, AND POLITICS

Roberto Barbanti, Isabelle Ginot, Makis Solomos, and Cécile Sorin

This book represents a milestone in a long-term project on the ecological turn in art carried out by a research collective based at Université Paris 8 (Vincennes Saint-Denis). This university is traditionally known for Philosophy (Gilles Deleuze, Jacques Rancière, and Jean-François Lyotard all taught here), Politics (Hélène Cixous founded the first gender and women's studies department in France here), and an anti-establishment culture that still thrives today. That the arts have been central to the creation of a revolutionary place for teaching and research since this experimental university was founded in 1969 is less well known. Today, nearly a quarter of students are enrolled on different arts pro-grammes. As the four editors of this volume, we teach and conduct research in the fields of Visual Arts (Roberto Barbanti), Dance (Isabelle Ginot), Music (Makis Solomos), and Cinema (Cécile Sorin). We have approached this book in a way that continues the tradition of the arts at Paris 8. At this university, arts are considered, not as objects of study, but as processes that produce thought within the artistic works and practices themselves. Our so-called "the-oretical" works are always anchored in practices, and we do not make categor-ical or essentialist distinctions between "theoretical research", "research creation", or "practice-based research". Artists play an active role in our research. Creation, teaching, and research are porous, overlapping categories. These different formats and disciplines come together to form: a robust research strand that combines a resolutely political approach to the arts and arts research; a belief in the emancipatory and subversive potential of the arts, research, and teaching; and a desire to disseminate knowledge beyond the academy.

This strong link to Paris 8's academic research and political culture belongs not only to the four editors but also to the wider collective that contributed to the production of this book. Over the years, this polymorphous group[1] has included members at Paris 8 and elsewhere, extending its reach far beyond the university. The collective is the real author of this work. Not only were all the contributors game to write "entries" that take the form of manifestos rather than traditional research articles, but some positions developed across this book were not evident at the outset and evolved through working with the collective. From the beginning, we intentionally adopted a rhizomatic methodology that led us to conceptualise this work, not with the comprehensiveness or systematic assimilation belonging to the logic of a dictionary or an encyclopaedia, but rather based on a desire for the openness and multiplicity that underpins an ABC Primer or "*abécédaire*".[2] This present work thus represents the sum of these experiences, encounters, and exchanges.

For over ten years, we have examined the relationship between artistic practices and ecological thought to explore the ecological turn in art. We have taught classes, delivered seminars, organised conferences and study days, and run experiments and labs; we have published texts and documents in print and online.[3] The subjects, research, and relationships have fertilised and taken root. There have been transplantations and transformations, international circulations, and extensions. We have created space for meetings and debates between researchers and artists, between artistic disciplines, between concepts and hypotheses, between generations, and between action, reflexivity, and analysis. We have expanded the methodologies of research to include those of artists. We have participated in sensory walks; we have paused our debates to listen to sonorous, visible, and tactile milieus. Together, we have taught and learnt how artistic practices treat ecological and, more precisely, ecosophic questions, and how those questions influence practices. We have entered into debates between aesthetics, art, and ecologies and between disciplinary, political, and institutional traditions. The interdisciplinary nature of ecological questions has made us rethink our own theoretical and artistic histories and create a dialogue with multiple researchers and artists.

It would be remiss of us not to mention the "milieu" in which we finalised this book. For a couple of years, COVID-19 turned the entire world upside down. As it did for the whole planet, it affected our ways of doing, thinking, and working together – and remotely. To assess fully the contradictory effects of the pandemic at this time would be impossible: on the one hand, transforming the economy and blocking the productivist machine finally proved possible; but, on the other hand, key sectors of the economy exploited the crisis in favour of increasing disparities and shamefully and dishonestly accumulating excessive wealth. On a political level, spontaneous forms of collaboration and solidarity emerged and spread around the world. Yet we also witnessed crackdowns on protests and the erasure of the collective memory of ongoing mobilisations; a prolonged state of emergency; and the apotheosis of biopolitical intrusion and control within society. Ultimately, science revealed its situated significance and how entangled its research objectives are with those of politics.

The relegation of the arts sector to the rank of "non-essential" activities exposed the same contradictions. Successive lockdowns decimated artists, particularly in countries where the arts have no state protection or those who largely depend on tours and productions. Whole sections of contemporary creation and culture in the most precarious countries were disappearing; numerous artists, who were no longer able to travel, no longer had access to any income. Who in privileged countries worried about this artistic massacre? For some people, the suspension of international travel with its large carbon footprint was a cause for celebration in the name of ecology. At the same time, creativity via a range of online platforms (Skype, Zoom, Whereby, multiple Padlet boards to ensure "pedagogical continuity") did not account for the ecological cost of this hyperconnectivity. And yet, successive instances of being confined to one's home produced, in addition to works about the pandemic, numerous calls to seize the opportunity to change practices, such as a festival for confined arts or forms of choreography that are listened to. Many lockdown customs and new artistic practices went viral, thereby redistributing what is free and what must be paid for, disseminating practices that can be done at home, and questioning in new ways how art can reach larger numbers of people. But again, we also saw the extent to which these new practices remain "confined" according to political, social, and aesthetic borders that have not drastically changed. The avant-garde addresses the avant-garde, activists speak to other activists, each is stuck inside their own network, whilst celebrities make money from the same innovations. Everyday during lockdown, we were treated to the TV broadcast of a well-known star stuttering through a song from the comfort of their home, which only served to reinforce the complete invisibility, in the eyes of the general public, of other artists whose post-pandemic future remains significantly compromised. Between the innovations brought about by the pandemic, the changes to social structures and how these transformations are subsequently being used to the benefit of the already privileged layers of society and the arts, all we can say is that the art sector faces the same contradictions as other areas.

We chose not to dedicate an entry to the pandemic, whose viral and social consequences continue to infect the planet, or to how the arts and this radically ecological crisis have had an impact on each other. Some of the deeply controversial aspects mentioned above have been widely debated, but not sufficiently enough to develop a thorough understanding of the situation. A proper analysis of the pandemic should consider the fundamental importance of cracking down on emotional and sensory experience. Since COVID-19 spreads through contact, breathing, meeting, and our relationships to one another, it is on this level that its impact has been and remains most widely felt. Such a study would need to consider, for example, how attention spans, turned towards screens to the detriment of the environment, have been transformed and how the accelerated instrumentalisation of the arts favours surveillance, big data, and cognitive capitalism. It will be necessary to explore how the pandemic, in some sense, served and continues to serve as a pretext for introducing official and legitimate bans on relationships, feeling, being together, and the vital energies that come from them. The pandemic made and makes ecosophic perspectives, where environmental issues combine with social and mental questions, all the more necessary and urgent. Nonetheless, the dynamics of the post-pandemic era, which we have now officially entered into, should not lead to a return to the old order, but rather to a radical transformation. Every day, the need for transformation becomes more and more pressing and more and more vital.

Notes

1 In addition to the book editors, this collective includes: Yann Aucompte (Université Paris 8), Clara Breteau (Université Paris 8), Ulysse del Ghingaro (Université Paris 8), Antoine Freychet (Université Paris 8, 1993–2022), Alice Gervais-Ragu (Université Paris 8), Gala Hernandez (Université Paris 8), Kostas Paparrigopoulos (Technical University of Crete, Greece), Carmen Pardo Salgado (University of Girona, Spain), and Lorraine Verner (École des Beaux-Arts de Versailles). We must also acknowledge the other contributors of this book and the students and the PhD candidates who participated in our study days, conferences, seminars, and classes.

2 An *abécédaire* recalls *L'Abécédaire de Gilles Deleuze* (*Gilles Deleuze's ABC Primer*), a documentary shot in 1988–89, consisting of an eight-hour series of interviews between Gilles Deleuze and Claire Parnet.

3 See our website: http://www.artstransitionsecologies.art

INTRODUCTION

*Roberto Barbanti, Isabelle Ginot, Makis Solomos, and
Cécile Sorin*

Key Concepts on the Relationship between the Arts and
the World

This book invites readers to deepen their knowledge of how aesthetic relations
and current artistic practices are intrinsically connected to the world. For the
first time in the history of humanity, we are conscious of the fact that any possibility of externality is illusionary: it is impossible to escape the ecological
crisis. Whilst the new space barons are eager to promote their excessive delusions of a planetary migration, we are very much bound to the Earth. Even if
some individuals can explore the planet's outer atmosphere, humanity as a
whole is destined to stay here, and aesthetic practices largely acknowledge this
impossibility of escaping the world.

If we consider the state of our world, even in a measured and moderate way,
the facts are terrifying. Social disparities are widening, becoming unbearable,
and depriving many of their basic well-being. At the same time, a generalised
state of control is becoming recklessly widespread in a way that is directly proportional to a shameful "lack of control" in which entire groups of people are
abandoned to the inhuman fate of a deadly indifference. All forms of surveillance and oppression are increasing; authoritarian regimes and dictatorships
are proliferating; one-party politics is triumphing; and the fascistic and fascist
ideologies, believed to have been confined to the history books, are re-emerging
within the same democratic systems that originally produced them. The crisis
of the Earth and living beings is coming to a head. It is becoming, literally,
unsustainable for many plant and animal species, humans included.

Even if this situation is a significant cause for concern, it remains distant,
detached. Processes of destruction occur within a structural inertia mixed with

DOI: 10.4324/9781003455523-1

a semblance of worry. Apprehension is maintained and dramatised just enough to keep the consumerist machine and its ideological and economic interests running. The suicidal defenders of the neoliberal *Homo faber* sing in unison at the top of their voices: "The show must go on."

Contemplating what will become of the world (or the *ecumene*) in its current state by remaining indifferent to the specific modes of production for artistic practices, aesthetic relationships, and the dynamics of the psychosocial representations that accompany them would be an error. The same would be true if we thought of the arts as isolated and separate within an ethereal universe based on a totalising and satisfying self-referentiality, or in their "autonomy", not dominated by exchanges and struggles with the world but conceptual, as a preconceived given, synonymous with the ivory tower. The relationship between the arts and the world is intrinsic in nature: if this relationship is lost, the arts cease to be what they are. Art is created in a given historic temporality and presents specific modalities of productive and sensory relations to the world.

In this desire to open up current practices and experiences to the world, we will maintain three guiding principles. First, thinking of aesthetic practices as separate from the world would be to consider the universe as compartmentalised and easily dissociable, in which any interaction is essentially viewed as an epiphenomenon. Current aesthetic practices are conscious of and, in their intentionality, evidence an awareness of their internality within the world. Second, it follows that the non-neutral experience of aesthetic realities is a conduit for engaging with social and political questions. Approaches to aesthetic practices not only acknowledge the world, but concretely deal with it. As such, the notion of political and social engagement becomes intrinsic to artistic activity and aesthetic relations. Third, the notion of the territory is central to these practices and confirms them as situated. These practices are constructed, produced, and experienced in a place, a temporality, an existentiality, and under specific conditions.

All socially and politically engaged art is a way of responding to or resisting an unconscious incapacity to react, the stupefied state of being subjugated and enchanted by neoliberalism. Indeed, this stupor significantly concerns the sensual dimension and aesthetics as much as on a speculative as a material level.

On the one hand, epistemological and philosophical modernity has relegated the sensory universe to the level of secondary qualities, as referring and belonging to a minor subject matter. This limiting of sensory experience to its subjectivity and disposing of its real cognitive capacity is the outcome of a long historic process, which has already been widely discussed in the academic literature: the move to demote aesthetic subjectivity is realised in neoliberalism's modes of domination. This significant delegitimisation of the aesthetic sphere frays the sensory fabric that connects us to the living world. Over time, the complex threads that attach and bind the body to an organic continuity became no longer perceptible and the power of ontological links was forgotten.

On the other hand, we are confronted with a large-scale offensive against sensory experience on a material level that assumes multiple forms. A sort of strange and furtive neutralisation of the sensorium, a "loss of senses" according to Ivan Illich's expression (2004a), decisively manifests itself on two intrinsically interrelated levels: the physical and the social. At the same time, systematic surveillance through the incessant spatial and temporal control of our lives, and particularly our sensorial lives, produces the conditions for widespread subjugation.

Furthermore, on a macroscopic scale, the primordial planetary phenomena, caused by human activity, remain largely inaccessible to our senses: it is impossible to detect the shift in our planet's rotational axis (Deng et al. 2021) brought about by the melting of glaciers, ocean acidification, and the rise in the concentration of carbon dioxide and some radioactive elements in our atmosphere.

On the macroscopic level of our everyday life, a form of structural and aesthetic hubris of planet-wide industry exists. Since the nineteenth century, humanity has dedicated itself to the mass production of disparate elements whose molecular, atomic, or informational nature is radically beyond our sensorium. A never-ending list of phantom elements penetrates and populates our living environments: synthetic chemistry, electromagnetic waves, nanomaterials, genetically modified products, irradiating substances. Despite being of utmost importance, this unprecedented phenomenon of going beyond the limits of human and, more generally, animal perception is relatively rarely analysed. We are no longer direct witnesses sensorially implicated in the future of the world that we are creating, or more accurately demolishing, on a daily basis. We overproduce on such enormous scales that our sensorium is incapable of detecting it.

Moreover, modes of production have become structurally aesthetic, if we can say that, as capitalism is currently established around processes that are essentially aesthetic in nature. Indeed, based on the structural necessity of constant production and accumulation, current modes of production can no longer simply produce goods since the market is already saturated: a need must be created for them. The system is set up to track and profile each person and make them an easy commercial target: the production of the self is freely available as a consumer commodity. This end is achieved through the almost absolute control of sensory conditions, made possible by communication and information technologies and, by consequence, through the instrumentalisation of aesthetic choices. The current economic trends that push us to be "inventive", "creative", "fun", "smart", and "green" can be analysed and, it seems to us, adequately explained within this framework.

The arts are also called upon to contribute. The notions of artistic capitalism or aesthetic capitalism (Assouly 2008; Lipovetsky and Serroy 2013) developed in sociology point to this underlying trend. We must not only denounce the fact that capitalism's mechanisms of reproduction have become deeply rooted in contemporary art's modes of existing and functioning,[1] but also question its impact on the arts. This mode of production concerns not only economic and property relations but goes as far as constituting an interrelated milieu that is at once physical-natural, material-artificial, and psycho-social. Whether elite or popular, intellectual or commercial, consensual or dissensual, the arts are fed, maintained, and subjected to this colossal machine that has made taste its unit of measurement, a targeted variable on an individual basis that is fully quantifiable in its empirically observed and statistically evaluated occurrences. The systematic control that is implemented and automatised from the modes of aesthetic relations and their manipulation discredits and erodes forms of representation from within.

In light of these assessments, this book is intended as a tool, a test, and the adoption of a position. It is a tool for those who position the notion of *aisthēsis* as a constitutive element of the human and living being. It is a tool that we offer to those who are worried about the world's destiny, the future of other earthlings, and our current condition as humans amongst living beings. It is a tool that we hope will be useful to those who are looking for sensual approaches other than that of instrumental rationality, and to those who hope for a form of reason that is sensitive to and capable of ethical decisions in the face of exploitations and abuses of gender and culture, racial oppression, and symbolic violence.

It is a test because what we propose is part of a common future that calls for each of us to offer their unique contribution and, at the same time, to position themselves within and be part of a common future, a common construction of this world, in this world. It is also a test of our own positions and experiences. We are conscious that their complexity is such that contradictions run through us and that, instead of concealing them, we must take them into consideration.

Finally, it is the adoption of a position as, over our years of research and supported by the collectives with whom we worked, we have attempted to *practise* in the field of arts and ecosophy a form of "situated knowledges" as defined by Donna Haraway:

> So, with many other feminists, I want to argue for a doctrine and practice of objectivity that privileges contestation, deconstruction, passionate construction, webbed connections, and hope for transformation of systems of knowledge and ways of seeing. But not just any partial perspective will do; we must be hostile to easy relativisms and holisms built out of summing and subsuming parts. "Passionate detachment" […] requires more than acknowledged and self-critical partiality. We are also bound to seek perspective from those points of view, which can never promise something quite extraordinary, that is, knowledge potent for constructing worlds less organized by axes of domination. From such a viewpoint, the unmarked category would *really* disappear – quite a difference from simply repeating a disappearing

act. The imaginary and the rational – the visionary and objective vision – hover close together.

(Haraway 1988: 584–5)

The phrase "the imaginary and the rational [...] hover close together" could very well describe the essence of research on the ecological turn in art. What unites the members of these collectives is a shared belief in socially and politically engaged art and research and in the production of simultaneously embodied and situated knowledges. These knowledges envisage imagination and creation as the common responsibility of artists, researchers, and activists, and that the production of knowledge and understanding of the world requires the intelligence of all. Rather than seeking consensus and "representing" the relationship between arts and ecosophy, we have sought different points of view, a networked production of knowledges that, according to Haraway's definition, practises listening as much as producing works, discourses, or concepts.

From the Ecological Turn to the Ecosophic Turn

Arts, Ecologies, Transitions: Concepts and Practices accompanies the significant developments emerging in the fields of arts and theoretical discourses on art. These developments reject confining art to the sphere of "civilisational surplus" and are in line with the questions arising from the ecological, economic, and social crises and the crisis of representation that we are experiencing. Environmental ecology is a common rallying point for artistic practices in transition. Environmental activists find counterparts in the artists who draw on the climate crisis, the threat hanging over biodiversity, or even local environmental causes and issues to create artistic processes that are often associated with militant action. That is why it is still relevant to discuss the ecological turn in art, despite it already having been extensively documented (Spaid 2002; Brown 2014; Ardenne 2019).

Arts, Ecologies, Transitions aims to both expand and redefine the notion of "ecology" in connection with the field of art. The environmental crisis that we are witnessing occurs on all levels: the destruction of nature is happening alongside harmful effects on society, on the planet's poorest populations, on the body, and on the mind. It is no longer possible to distinguish between nature and culture – as ecological, philosophical, anthropological, and sociological research have recently shown. For example, Catherine Larrère and Raphaël Larrère (2015: 6) state:

The idea of nature, traditionally based on a series of oppositions (natural/artificial, savage/domestic, nature/culture, etc.), has [...] been called into question with environmental problems becoming widespread. What characterises the environmental question is its erasure of the division between the natural and the social on which western representation of the world (and especially that of modernity) was organised: on the one hand, there is what concerns nature (which scientists are responsible for); on the other, what is relevant for society (political matters and common life). In the environmental crisis, the natural trespasses on the social (our environmental problems are social problems) and the social leaves its mark on the natural (some of our social activities, particularly since the industrial age, damage nature).

Alongside environmental ecology, and largely intertwined with it, is *social* ecology. As Murray Bookchin stated decades ago:

> Social ecology is based on the conviction that nearly all of our present ecological problems originate in deep-seated social problems. It follows, from this view, that these ecological problems cannot be understood, let alone solved, without a careful understanding of our existing society and the irrationalities that dominate it.
>
> *(Bookchin 2007: 19)*

In addition, *mental* ecology designates the close relationship between the environmental crisis and the problems that individuals recognise through their feelings or ways of existing in the world.

Mentioning the three ecologies brings us to Félix Guattari (2000, 2013), a theorist whose idea of there being multiple, inseparable ecologies is eminently relevant to the current situation. The philosopher and psychoanalyst designates three "ecological registers (the environment, social relations and human subjectivity)" and envisages an "ethico-political articulation" between them that he opposes to the "purely technocratic perspective" with which politics tends to respond to the environmental emergency (Guattari 2000: 28). He names this articulation "ecosophy" (2000: 28), a term used also by another pioneer of political ecology, Arne Næss (1973). In this collective work, we attempt to bring together questions of *mental* ecology and *social* ecology with environmental issues to identify the different modalities by which art today redefines the processes of subjectivation and the emergence of collectives, challenges affects or relationships to the "body", focuses on listening, and interrogates the notion of the author, amongst other questions. We are therefore attentive to the ecosophic nature of artistic practices in transition, practices that are situated and engaged with environmental, social, and mental questions. We postulate that ecosophic practices exist in art, even if they are not claimed as such by the artists themselves.

This ecosophic approach to artistic practices requires other conceptual decentrings. Arts in transition become ecosophic when they question the very notion of *aesthetics*, demanding a return to its etymological roots of *aísthēsis*, sensation, perception, a practice of feeling. This connection between the three ecologies thus plays out and can be captured in the notion of *aísthēsis*. From this displacement, new objects, projects, processes, and situations emerge to explore relationships to space and the environment, different forms of temporality, the contiguities with citizen action or even experiences of everyday life: performances that interact with living beings and technologies, art or sound walks, productions with performers that have disabilities, artistic practices with non-traditional audiences, experiences of "deep listening" and somatic listening, and documentary arts. At the same time, a new theoretical vocabulary has replenished and regenerated research aesthetics in recent decades. This vocabulary is not only evidence of a conceptual revival, but it has contributed to its very formation. With the concepts of aesthetic experience, ambiance, atmosphere, landscape, environmental and ecological aesthetics, sensory relationships, and the sensual turn, we move away from, while encompassing it, the idealist philosophical tradition that exclusively defined aesthetics as a philosophy of art.

Likewise, arts in transition adds an *ethical* dimension to aesthetics. The passage from aesthetics to ethics sometimes suggests a "po-ethical" or even an "aesth-ethical" dimension of art. Today, the revived ethical question is asked through the artistic practices that conceptualise or intervene on the level of the organic and inorganic world, of "bodies" and techno-sciences, often going beyond representative modalities and placing itself directly in the real, all the while establishing the conditions and modes of artistic production. Art opens up to address the complex questions that define our world. The concepts of limitations, justice, responsibility, or inequality are implicated in this examination, in other words all that reveals the pragmatic universe.

Whilst the term "ecology" remains largely attached to environmental ecology, the artistic practices and the artists discussed in this book (e.g., those who work on questions of society, gender, class, or race without necessarily using the lens of environmental ecology) do not always mention ecology. Likewise, given our focus on practices, the ecological dimension does not always appear as an explicit theme. Ecological, and more broadly ecosophic, art is not necessarily a type of art that focuses on melting ice caps or climate change. However, it must not contribute to the climate crisis, and it must have the means to sustain the struggle in the long term through the forms that its very existence takes.

The idea of *transition*, which is well known in the sphere of political ecology, makes it possible to address these developments, which are not the same as the transformations or ruptures that are usually studied in modern or postmodern art. Transformations or ruptures are more formal in nature, whereas the transitions discussed here often require a redefinition of the notion of art itself. As Rob Hopkins, the founder of the Transition Network, writes:

> Transition is one manifestation of the idea that local action can change the world; just one attempt to create a context in which the practical solutions we need can flourish. [...] Transition is an idea about the future, an optimistic, practical idea. And it's a movement you can join. There are people near you who are optimistic and practical too. And it's something you can actually do. Actually, it's lots of things you can actually do. Lots of things. The Transition approach is self-organising and people-led; it looks different everywhere it emerges, yet it is recognisably Transition.
>
> *(Hopkins 2013: 48)*

In this line of thinking, the political movements that wish to change the world should not wait for the revolution, the new dawn to come. However, it is important not to confuse the idea of transition with that of classical reformism. The ecological crisis (that of environmental and social ecology) cannot allow for half measures. Its goal cannot be: "Fight against pollution and resource depletion. Central objective: the health and affluence of people in the developed countries" (Næss 1973: 95). Even in 1973, Næss criticised this approach as a "shallow ecology", opposing it with a "deep ecology", characterised by, amongst other things, "rejection of the man-in-environment image in favour of *the relational, total-field image*", "*biospherical egalitarianism*", and the "*principles of diversity and of symbiosis*" (1973: 95–6, italics in original).

Arts in transition, linked to ecological and ecosophic questions, do not "address" environmental, social, or mental issues; indeed, they resist the logic of greenwashing. Can you imagine an artist tackling rising sea levels whilst being financed by the oil and gas giant Total? Arts in transition take up environmental, social, mental, political, or gender issues to create artistic questions, which transforms the very essence of art itself. Art thus becomes a milieu in which a new ecosophic consciousness is forged. Some artists pass over into action, giving rise to forms of artivism.

In addition, the notions of "work" and "author", which have not lost their historical relevance and their contemporaneity, are opened up to the complex world that contains them. Based on this, the finished work is reopened to the temporalities of its becoming and the conditions of its existence, thus giving form to notions such as those of process, medium, interrelation, and contextuality. The artist-genius multiplies in the sensual movements that populate the social and allow for their emergence. The self-centred ego must negotiate its narcissism in the collective processes of subjectivation, whilst the pretence of a resolutely anthropocentric vision is exposed to the feelings of other living beings.

As for this book, we do not privilege modes of expression, movements, or even artists or works over others. Nor do we define or evidence a supposedly ecological, or even ecosophic, current in art. What interests us is presenting the heterogenous proliferation and extension of aesthetic practices that uniquely characterise this historic phase: the sensory awareness that they bring about; the changes that they manifest or announce; the contradictions that they express; and the levels of reality that they reveal. In short, we are interested in the complexity that aesthetic practices convey, their capacity to read reality and voice the emotional issues of our time and in our time, the common and the intimate.

Thus, we have explored the transition from the anthropo-ego-centric paradigm of man, the subject-artist and the work, towards the conditions and limitations of artistic possibility, shared feelings, the non-commensurable richness of aesthetic experiences and their organisation in a co-evolutive becoming, that is to say in common and co-constructed with Others.

To do this, we had to develop a protocol for collective working that drew on the principle of co-construction in line with Guattarian thinking. Our aims were to make the friction between different subjectivities visible, to produce a rhizomatic book, and to valorise the very processes of development and writing. Our work began with organising seminars and study days, which led us to establish the list of entries and contributors. Each decision was taken collectively as part of the free exchange of ideas that this unique time required and which is inherent to any experimentation process in which thought, formulated together, emerges.

A Manifesto

The question of "Why is Guattari so rarely cited in Green Studies?" is emblematic of the major differences in ecological approaches to the arts in the anglophone and francophone academic worlds. Previous studies on these differences (see Posthumus 2010; Suberchicot 2012; Vignola 2017) tend to underline the specificity and the complementarity of these approaches that, in broad brush strokes, limit anglophone Green Studies to an evaluation of the discursive efficacy of works, whilst the francophone, mainly ecopoetic, approaches principally deal with what constitutes the artistic functioning of works.

The notion of realism is emblematic of the ontological differences that separate these two approaches: "the influence of the French tradition – the age of suspicion, the *Nouveaux Romanciers*, the writings of Deleuze, Derrida, Foucault and Lacan amongst others – makes the return to a simple realism impossible" (Posthumus 2010: 148–9). Approaches to representation that are widely used in cultural studies would analyse the memetic relationship between works and nature and evaluate the gap that separates them, as if inhabiting the *oikos* should necessarily produce works of heightened, fusional, or militant realism.

Ecology in art cannot limit itself to the discursive, formal, or strictly representational level. Instead, it should call upon the sensory dimension of the work for several reasons. First, ecological awareness, like the ecological lens, passes through an intimate, physical experience (Lussier 2011: 257): ultimately, what is described as an argument or a discourse has little impact in terms of political effectiveness. Second, following the example of space, we experience the world; it can only be partially described. To experience the world through works of art operates on the registers of the senses and action, more than that of discourse. Third, in Guattari's *The Three Ecologies*, the production of works and their reception are interlinked within a social, ideological, economic, ecological, cultural, and technological context. If the level of overlap is of such a complexity that it is difficult to describe, it is nevertheless possible to feel it precisely due to the sensual experience of the work and the analysis of the processes of subjectivation.

Finally, and perhaps above all, there are numerous aesthetic practices where the ecosophic dimension resides in the restoration, for the artist and for their audiences or for participants, of what it is necessary to call an *experience of nature*. In other words, this means not only transmitting knowledge about nature and the violations that are weakening it, but also inviting audiences to experiences that reconnect with the feeling of being part of nature. This sentiment is largely eclipsed by the modalities of urban life and the modes of subjectivations inherent in neo-capitalism.

To mark a milestone in the process of collective research, we decided to propose complementary approaches to cultural studies that clarify the multiple relationships between art and ecology and to conceptualise the book as an *abécédaire* of key concepts. It is not an encyclopaedia. It does not establish the current status of knowledge on ever-evolving concepts, nor does it offer a systematic and exhaustive sweep of the totality of this field with its infinite ramifications. At certain points, this book deliberately distances itself from academic discourse to give space to other artistic or socially and politically engaged forms of communication whose plurality sketches out the vitality of reflection and creation within sensual approaches to ecological questions.

Some detours through notions related to the fields of art and ecology, without addressing them head-on, seemed necessary. It is a case of contextualising the different elements of critical thinking and at best situating the key issues of these debates within art and activism. Faced with the urgency to consider the "collapse" in its social and cultural complexity, it was essential to leave space for more direct and dynamic forms of address. This book thus assumes its function as a manifesto through the form of certain entries and the promotion of ecosophic thinking.

This inherent polyphony is also disciplinary. The book maps how ecosophic questions are embodied when rooted in different disciplines. Each entry's author speaks from the perspective of their discipline, their field of research, or their own area of art. All the artistic fields (visual arts, theatre, dance, music and sound art, cinema, photography), including those that are rarely represented in the academic world (e.g., digital creation or graphic design), are addressed in a way that underlines the diversity of artistic practices in transition.

The focus is on practices more than works. There are no entries on artists, works, or theorists. By renouncing this, we reject the dominant method modelled on a discourse that positions the artist and their works as uniquely legitimate. Artists and works, however, are not absent from this book. On the contrary, they are present and disseminated amongst the entries. Likewise, some concepts, like "catastrophe" or "plants", do not have their own entry because they are addressed in an iterative way through different entries and viewed from complementary angles and approaches that emphasise their polysemy and reveal the plurality of certain issues.

Bringing everything together around an ecosophic, Guattarian approach, far from producing unity or a consensus, allows us to seek variations and divergences that make some artistic practices ecosophic practices, most often in a partial way. Some entries favour a local scale, others a planetary scale; some focus on the dynamics of subjectivation, others on the environmental or social dimension. At all stages of this process, we sought to highlight the scope of debate and different points of view, to promote diversity in terms of formats, lenses, and approaches. The same concern for diversity and debate drove the call for contributors. What is common to all contributors is the political engagement of our work. We issued a call for subject specialists who share our collective's commitment to social and political engagement, without looking for consensus. Likewise, the tone, the mode of writing, and the style of each entry varies from manifesto to academic article.

This book does not seek to produce general knowledge about the arts and ecology, rather it brings together a plurality of situated discourses that are partial and rooted in different situations. The concepts of listening, perception, immersion, and tactility, which are constant sources of inspiration for artists who think in *terms of milieus*, are also at work in our bodies as researchers, which provides an example of how Guattari's mental ecology is embodied through artistic practices. It also allows us to understand how the individual dynamic of subjectivation links with the collective and the social. *In situ*, participatory, or collaborative artistic practices, even if they are not identified with social ecology, multiply in an infinite diversity of propositions to create bonds, inhabit a social space, and reinvent the modalities of being together. Artistic practices often begin from a "doing together", which offers a reading of natural and urban milieus, the modulations of our relationships to objects and to our environments, and, to a certain extent, the invention of degrowth. In other words, if no artistic practice can claim to cover the totality of the issues in an ideal ecosophy, all of them experiment, on their own scale, with modes of articulation across three levels: the environmental, social, and mental.

Finally, as editors of this book, we would like to end this short introduction by acknowledging and, for both the readers and the contributors themselves, shedding some light on the double complexity that constantly enriched our work. We sought to conduct the research relating to this book through shared suggestions and decision-making. Given it formed the basis of our approach, it also had to apply to our editorial work. The task of collectively writing in English for an international audience posed intellectual and cultural challenges that manifested themselves in the process of translation. How, for example, to translate the quality of being "*intraitable*"? On the surface, it suggests "that which cannot be treated", but "*intraitable*" assumes an additional, complex dimension when it appears in a quotation by the Martiniquan writer Édouard Glissant. In this instance, "*intraitable*" also references the "*traite des Noirs*", the slave trade, and the way in which it has been "*intraitable*", obscured and downplayed by the colonial French state to the extent where there are no words to express what has been lived and what has been silenced. Similarly, with Estelle Zhong Mengual conceptualising the notion of "*arts en commun*" as distinct from the commons, using an existing term in English for socially engaged artistic practices would not have accounted for this difference. Capturing these types of subtleties and nuances throughout the entries threw up many tricky translation challenges. Added to the inherent difficulties of transposing ideas from French to English for a broad, international readership were issues relating to the multiple thematic approaches adopted and disciplinary-specific terminology in each contribution. Each entry was read by the editorial team and discussed with its author before being translated, then reread by the editors and the authors, after which it was finally approved. To achieve a sense of stylistic consistency and clarity, the considerable task of translation was entrusted to one person, and we would like to thank our translator Stacie Allan for her expertise, precision, and professionalism. All translations, unless otherwise stated, are by Stacie Allan.

Whilst this two-part process of writing and translation opened up a very enriching path, the desire to honour the complexity of the task means it has taken time. In a field like research where speed and competition appear to be the requisite motives behind every individual (and, as it happens, individualist) choice, we took a gamble by adopting a different temporality and other relational and decisional modalities: those of a disciplinary consciousness that is at once multiple, expanded, and shared. Even though the deadlines gradually stacked up and, now and then, the labour-intensive work discouraged us, we can today appreciate the validity and richness of this approach.

We would like to thank everyone who participated in this project for their expertise, engagement, and openness.

Note

1 The artist works all day every day: creativity does not take a break. The artist radiates individualism and the cult of personality. Their signature is the proof of their absolute uniqueness. The artist fully embodies their work. They express their whole being in their work; they accomplish themselves in its production. The artist exists in and for their creations. They perpetuate the ideal of eternal competition: they want to establish their brand; they establish themselves through widespread competition.

1

ACOUSTIC ECOLOGY

Kostas Paparrigopoulos

What is Acoustic Ecology?

Acoustic ecology is a discipline that studies the ecological aspects of sound, the relationship between living beings and their sonorous environment. The term groups together several scientific and/or artistic trends that view sound as a phenomenon in and of itself and, above all, through its relational aspects, often drawing on interdisciplinary approaches. Other disciplines and approaches offer similar terms to acoustic ecology, such as sound ecology, soundscape studies, soundscape ecology, ecoacoustics, or even sonorous ecology (see Barbanti 2011).

From the late 1960s onwards, the Canadian composer Raymond Murray Schafer, who invented the concept of the soundscape, produced innovative studies that theorise approaches to an environment's sounds. In his seminal text *The Soundscape: Our Sonic Environment and the Tuning of the World* (1977), he calls for a revaluation of soundscapes, from the mediocrity of a *lo-fi* soundscape, filled with loud noises that mask the quietest sounds, to a rich *hi-fi* soundscape where all or almost all the sounds are perceptible. Schafer views the world using the enchanting metaphor of an enormous musical composition, without an end and probably, as he suggests, without a beginning. This composition is happening around us and it is our responsibility to enhance or destroy it (Schafer 2009). Schafer thus proposes "ear cleaning" exercises and "soundwalks" that can "reopen" our ears and "rediscover" the sonorous environment (Schafer 1977).

Lo-fi soundscapes are primarily associated with the noisy city and hi-fi soundscapes with the peace of nature. Hi-fi soundscapes become a sort of acoustic ideal to aim for, which can also contribute to the fight against sound

DOI: 10.4324/9781003455523-2

pollution in cities. Following this line of thinking, the ideal city would resemble, acoustically speaking, the countryside (for criticism of the hi-fi/lo-fi opposition, see Paparrigopoulos 2016b).

Schizophonia

One of the concepts fundamental to acoustic ecology is schizophonia. This term is used to describe the separation (schizo) of a sound (voice – phonia) from its source, the temporal and spatial context where it is present. This separation, made possible with sound recording, facilitates the electroacoustic reproduction of a sound in other moments or places (*ex situ*), which Schafer considers as a negation and degradation of its natural sound: "I employ this 'nervous' word in order to dramatize the aberrational effect of this twentieth-century development" (Schafer 1977: 273). Despite Schafer's criticism of "electric" sound (essentially due to its omnipresence in everyday life and its often excessive volume), schizophonia is a very useful concept in sound research as well as in artistic expression inspired by acoustic ecology.

Acoustic and Musical Ecologies

By defining the concept of a soundscape, Schafer simultaneously opened up avenues for music and for soundscape compositions. On soundscapes, he writes, "the term may refer to actual environments, or to abstract constructions such as musical compositions and tape montages, particularly when considered as an environment" (Schafer 1977: 274–5). Researchers at Simon Fraser University theorise compositions, based on soundscapes and linked to acoustic ecologies, in a way that aims for "the re-integration of the listener with the environment in a balanced ecological relationship" (Truax 2008: 103–9).

Compositions based on soundscapes can be considered as a branch of electroacoustic music that also promotes an ecological ethic. According to Barry Truax, one of Schafer's closest collaborators, the principles of soundscape compositions are as follows:

> Listener recognizability of the source material is maintained; the listener's knowledge of the environmental and psychological context is invoked; the composer's knowledge of the environmental and psychological context influences the shape of the composition at every level; the work enhances our understanding of the world and its influence carries over into everyday perceptual habits.
>
> *(Truax online)*

Moreover, environmental sounds are being introduced into contemporary forms of artistic expression – which, traditionally, were not sonorous – and often through an ecological approach, such as in performances, installations, or *in situ* events. This increasing interest in sound (not only in "musical" sounds) has characterised musical practice since at least the start of the twentieth century (see Solomos 2013).

Field Recording

Field recording or phonography consists of recording the sounds of a place, often in nature and outside of a studio, on an analogue or digital medium with the aim of reproducing and transforming them (soundscape compositions, sound art, etc.) or documenting and archiving them. Bernie Krause (1987) is known for his field recordings in relation to bioacoustics (biology-acoustics). In the humanities, field recording has always been used in enthnomusicology, as well as also being present in disciplines like zoomusicology (animal music) introduced by François B. Mâche (1983) or biomusicology (biology-music) introduced by Nils L. Wallin (1991).

Acoustic Ecology and Transition

Acoustic ecology has opened up areas of research, reflections, and practices that have quickly been embraced by different scientific and artistic disciplines. The transitive nature of contemporary societies, which pay attention to the sounds of everyday life, facilitates the development of an auditive, social, and environmental awareness. According to social ecology, researching a balanced soundscape presupposes the existence of a similar social organisation (see Bookchin 1993). Correspondingly, an unbalanced society deserves its own analogue soundscape. We might then ask: What role can acoustic ecology and its artistic expressions play in the search for equilibrium? Hildegard Westerkamp, pioneer of acoustic ecology and a highly innovative soundscape composer, writes:

> Can soundscape composition initiate ecological change? This is the challenging question to all of us, whether soundscape composer or soundscape listener. Can we become active acoustic ecologists no matter whether we create the compositions or whether we listen to them? Isn't it precisely in the link between composer and audience that energy for change can be created? And isn't it precisely in the link between soundscape composition and acoustic ecology that meaning is created? Here cultural production can speak with a potentially powerful voice about one of the most urgent issues we face in this stage of the world's life: the ecological balance of our planet. The soundscape makes these issues audible. We simply have to learn to hear it and to speak back. The soundscape composer has the skill and the expertise to do exactly that.
>
> *(Westerkamp 2002: 56)*

Three Key References

Schafer, Raymond Murray (1977). *The Soundscape, The Tuning of the World*. Vermont: Destiny Books.

Truax, Barry (2008). "Soundscape composition as global music: Electroacoustic music as soundscape". *Organised Sound* 13(2): 103–9.

Westerkamp, Hildegard (2002). "Linking soundscape composition and acoustic ecology". *Organised Sound* 7(1): 51–56.

2

AESTHETIC SUBJECTIVATION (*PROCESS OF*)

Roberto Barbanti

Following on from Galileo's epistemological modernity, René Descartes established the subject in an anti-aesthetic duality in the seventeenth century. The mind was ontologically separated from the body, the quantitative dimension from the qualitative, truth from the capacity to feel. This Galilean–Cartesian concept of the subject would forge modernity and dominate thought de facto until the second half of the twentieth century.

Drawing on a renewed aesthetic consciousness, Michel Foucault challenged the modern genealogy of the subject through the lens of the power relations that it establishes between humans. In 1982, he wrote: "There are two meanings of the word 'subject': subject to someone else by control and dependence; and tied to his own identity by a conscience or self-knowledge. Both meanings suggest a form of power which subjugates and makes subject to" (Foucault 1982: 212). Power thus operates both as external coercion and internally. The subject is returned to the centre of analysis. However, this time, it is no longer the "pure" and "transcendental" subject of modernity, but a subject in the process of becoming, a living being of flesh and feelings, historically determined and contextualised, whose designation would lead Foucault to rethink the conceptions of "subjectivity" and "subjectivation". The French philosopher advocates for an aesthetics of existence, a capacity for autopoiesis that is not concerned with isolated, abstract individuals, but concrete singularities caught within a common becoming.

This broader conceptual understanding of the subject as an "I" within a sensorial and thus aesthetic "we" would be taken up again by Félix Guattari. Following Foucault, Guattari showed the decisive importance of subjectivity whilst further expanding the theoretical concept. In 1992, he stated: "[subjectivity] has become the number one objective of capitalist society, of

DOI: 10.4324/9781003455523-3

contemporary society" (Guattari 2011: 40). Yet, it is not a question of valorising the mode of production that conquered the world. Much to the contrary, "subjectivity produced on the industrial scale of the mass-media is a reduced, flattened, devastated subjectivity, which is losing its singularity" (2011: 42). Faced with subjugation, and this flattening of singularities, Guattari proposes a shared and conscious capacity to construct a common becoming, understood as a process of collectively assembling subjectivities on the basis "of a new aesthetic-political paradigm" (2011: 49). In other words, it is the creation of "new modalities of subjectivity in the same way that an artist creates new forms from the palette" (Guattari 1995: 7). Moreover, Guattari's notion of ecosophy places the question of the subjectivation process in an expanded horizon that "is completely beyond the anthropological sphere" (Guattari 2011: 41), taking into account not only the relationship to oneself and to the other (*socius*), but the environmental context and other species as well.

The inclusion of non-human living beings as social beings who have their own interests breaks "with the idea that humans alone make history" (Stengers 2019: 22). As Isabelle Stengers writes, "less and less are we able to talk about human histories without including their adventures with their environments, the manner in which they transform them and are transformed by them" (2019: 22). With reference to the concept of sympoiesis proposed by Donna Haraway, Stengers insists on the necessity of "making *with*" (2019: 28) other species. Furthermore, the Brazilian anthropologists Déborah Danowski and Eduardo Viveiros De Castro affirm that "we call those for whom the world has always been the world or, better still, *of whom* the world has always been the world, 'humanity', 'humans', or even 'us'" (Danowski and De Castro 2014: 234). In other words, "us" represents the beings who can perceive the world and to whom the world belongs. The concept therefore is not a reduced or homogenised "we" that is indifferent to all diversity, nor another false new subject in an immaterial and anti-aesthetic abstraction. It is not a disembodied "we" whose thoughts and feelings are not rooted wherever it finds itself. In contrast, this "we" is conscious of its profound relationship to *aísthēsis* and thereafter convinced of the impossibility of perpetuating the idea of subjugation. This "we" seeks to decolonise imaginaries and practices.

This notion of subjectivity is concrete and multiple: in the act of becoming, it possesses an inexhaustible multitude of singularities linked to a common destiny. This "we" brings together the Earth's inhabitants through their capacity for sensorial experiences, all as shared as they are varied and complementary, and creates as many worlds as unique interrelations between worlds. It is an aesthetic subjectivity that opposes both the deaestheticised modern subject and the persistence of an aesthetic capitalism that reifies, controls, produces, and alienates the sensorial subject through a widespread fetishisation of commodities. Challenging the process of subjectivation from the notion of aesthetic subjectivity thus poses the question of a new materialism as a surpassing of the ethereal subject that is overly rationalised and intentionally "an-aestheticised".

It is thus necessary to begin with the materiality of *aisthēsis*: the connected, shared, and heterogenous form of perception that produces singularities entwined with earthly links. However, we must first consider the historical specificity of Western *aisthēsis*'s Greco-Roman heritage, which created a sensorial hierarchy that privileges the visual dimension. Its theoretical becoming – theory: *theôrós*, the "spectator" – promoted cognitive modalities that are disjunctive and decontextualising, abstract and linear in our relationship to the world. In the act of assigning meaning, "the Western gaze" gives priority to the static dimension of space removing all temporality from observed objects to fix them in an ontologising and essentialist vision. These tendencies were reinforced and lauded by modernity and integrated into nineteenth-century German thought, which advanced the idea (Latin *idea*, from the Greek ἰδέα), understood as the "visible form", potentially to its purest degree of abstraction. Indeed, in the anti-aesthetic variants of the Hegelian matrix, *Weltanschauung* has accustomed us to a conceptual reductionism that drives the act of feeling exclusively to concrete manifestations of the spirit, that is artistic production. The work, the author, the genius were made absolute as the only realities worthy of interest, thus strengthening and confirming the modern separation between the "subject" and the world. This way of thinking decontextualised the arts and placed them in a self-fulfilling mythology of the absolute, leading to a delegitimisation of forms of feeling that do not directly lead to what is recognised as productive will. This conceptual, agentive, and normative paradigm which continues to be perpetuated has been amply challenged by a set of intentional aesthetic experiences. Through these experiences, an innumerable number of actors have brought about greater perceptive awareness across morphological research practices that thematise and experiment with the role of *aisthēsis* in the world. Collaborating with the living world and studying its languages; reclaiming the body and reformulating dominant historical narratives; walking, making "derives",[1] and travelling across landscapes; and critical immersion and subverting technical environments are but some of the modalities, amongst so many others, that, by refusing subjugation, construct a mode of relation to oneself within aesth-ethical and po-ethical commons.

Emerging from the dialogic intersection (Morin 1990) between milieus and singularities as well as establishing a new *aisthēsis*, the concept of subjectivation today is developing as a process of production of aesthetic subjectivity and raising the question of a new aesthetic materialism that goes beyond the spectator-subject, disconnected from a wealth of reasoning and bewitched by the laws of commodity fetishisation.

Note

1 The situationist theorist Guy Debord defines the "derive" as "a technique of rapid passage through varied ambiances". See https://www.cddc.vt.edu/sionline/si/theory. html

Three Key References

Abram, David (1996). *The Spell of the Sensuous: Perception and Language in a More-Than-Human World.* New York: Pantheon Books.

Escobar, Arturo (2014). *Sentipensar con la tierra: Nuevas lecturas sobre desarrollo, territorio y diferencia.* Medellín: UNAULA.

Pignocchi, Alessandro (2019). *La recomposition des mondes.* Paris: Seuil.

3

AISTHESIS

Carmen Pardo Salgado

Building a common home.

Our experience of sharing a common world has been completely disrupted and our place on the planet is beginning to be contested.

At a time when common sense, understood as a common ability to feel or perceive, is subject to transformation, it is becoming necessary to re-examine the concept of *aisthesis* as a tool for reflecting on and opening up other ways of taking action.

We must seek another common form of sensing, a shared capacity to feel that moves away from conditioning both thought and bodies. We must seek a form of sensing and common sense that is not complicit in the demands that subject the planet to financial madness.

The time has come to create a political experience of our senses, to create an event of the senses.

The time has come to question how we share our senses: forms of seeing, hearing, touching, tasting, and smelling; forms that anchor the body in ways of thinking.

What do I look at the most? What and how do I see? What am I receptive to touching or not touching? How do smells reach me? What types of listening do I use with other humans? With the sounds in my environment? With those who emit these sounds? How do I work on my taste for things, others, and everyday life?

Creating an event of the senses means uncoupling them from the apparatuses of serialisation that underpins sensorial perception and the senses, transforming all organic and inorganic beings into a reserve to exploit. These apparatuses are built on behavioural practices that orientate sight, touch,

DOI: 10.4324/9781003455523-4

smell, listening, and taste. We are all influenced by these apparatuses through our concrete relationships with the environment, ourselves, and others.

Creating a political experience, an event of our senses supposes redrawing the intimate relationship between the educational milieu, questions of gender, marginality, poverty and wealth, and power, as well as with ourselves.

Creating a political experience of our senses means understanding that, whilst Kant's *aisthesis* in the Age of Enlightenment was linked to a critique of taste, the concept today will be rooted in a practice of tactility.

We need more than a critique; we lack practices that place a reciprocity of touch at their centre and remind us that touch draws on all the senses.

Examining *aisthesis* at this present time means prioritising the question of practices.

A practice is an action in which sensory experience and the senses are embedded and work on and with what exists.

It is thus a political experience of the senses, an event of the senses that comprises an ability to consider what makes us move, touch, and set things in motion in order to get closer to what we have difficulty understanding: the desire that makes us act.

Desire is the strength, the inherent power in the formation of sensorial perception, that leads to tactility, that weaves relations between all that is perceived and those who perceive. This relationship does not take one single form: I perceive and I am perceived.

To desire is to make, to form relationships that can change the environment and transform social and mental bonds.

These relations construct the complexity of the sensory event.

The *aisthesis* on which these relationships are, at present, constructed consists of an appropriation whose watchword is the instrumentalisation of all living beings. Its driving force is the way in which any desire – even any non-desire – is made available to the urge for possession, for infinite accumulation. Desire is conceptualised as lack and all actions must be directed towards satisfying that lack.

Desire is equated with a desire for possession. I have eyes, ears, a nose, a tongue, legs, skin: I have a body. And thus, it continues: I have a family, friends, neighbours, a dog, a cat, an apartment, a house, plants, a car. I have. I have. I have.

What happens if we replace "I have" with "I am"?

I am: eyes, ears, nose, tongue, legs, skin … body.
I am: family, friend, neighbour, dog, cat, plant, mineral, apartment, house, car.

In each instance, what "I am" must change, transform itself, adopting different forms, movements, and behaviours according to who or what it becomes.

Desire is not only a lack. Desire is the inherent power of a body that thinks and feels. Desire is the fundamental relationship to the world and to oneself. Centring desire on a lack or an absence means forgetting that desire is a power that determines the field of action, whether that field is already prepared or to be created.

Desire is a passage across which senses and thought interweave. Desire is our porosity. But, before and after, there is no "I" who wants to have or feel. There is an *us*, a *one*, an *us all*.

How do we move away from lack or absence and instead place emphasis on power? How do we uncouple the desire from the framework that feeds this infernal system? What practices can be developed?

To uncouple desire from this framework, we must work on relationships; we must be preoccupied by them, be their *bricoleur*.

The first task of a *bricoleur* of relations consists in applying their practices to understand the relationship between one's desire and absence, possession, true and false, joy and sadness, others, objects, and the beings that accompany it.

The time has come to examine the relationships between our desires and the apparatus that structure meaning, our senses, and our shared capacity to feel: media, educational and health institutions, leisure activities, work.

It is not about blaming or demonising. The time for complaining is over.

The time has come to recognise our affiliations and our responsibility in this global laboratory.

The time has come to be able to think and feel on a micro-, as well as planetary, scale.

Reconsidering *aisthesis*, basing it on practices, leads to the necessity of transforming an ecosystem that often forgets its relational character.

Considering *aisthesis* today draws on an ecology understood as a web of multiple interactions between habits and behaviours, sometimes ordered and at other times chaotic, between the environmental, the social, and the mental.

The experience of a common world, a world in common, comprises disagreements and unusual agreements, polyphonic pronouncements and fragile articulations.

The experience of a world in common means learning to interact with each other in different ways, wanting to build in straw and iron the practices, at once flexible and robust, that construct the home we share.

Three Key References

Dardot, Pierre, and Christian Laval (2014). *Commun. Essai sur la révolution au XXe siècle*. Paris: La Découverte.

Kostelanetz, Richard (ed) (1988). "Esthetics", in *Conversing with Cage*, Richard Kostelanetz (ed) New York: Limelight Editions.

Rancière, Jacques (2011). *Aisthesis. Scenes from the Aesthetic Regime of Art*. London/ New York: Verso.

4

ALIENATION

Eric Lecerf

Amongst the numerous contributions that Gilles Deleuze and Félix Guattari made to philosophy, their criticism of doctrinaire usages of "concepts", which delimit life, must be acknowledged. It is at "the point of one's ignorance" (Deleuze 2012) that the individual resorts to a concept. The rest of the time, life in and of itself is enough to solve life. The *moment of the concept* is thus an insistence of liberty that sees the subject develop in relation to a world where the unique has its place. To continue to maintain a direct relationship with the dynamic flux that gives life its entire meaning, the concept becomes enriched through heterogeneous "articulations of the real" (Bergson 1946: 58). This is the only way that the concept can continue to question its meaning, to resonate in contact with life. Fixing any concept according to a strict definition must therefore be avoided. Inspired as much by Bergson as by Saussure, Deleuze and Guattari remind us that a concept is conceived within a chain of meanings in which no single word has meaning in and of itself. This offers an entry point to understanding Guattari's three ecologies, three frameworks to make sense of the world. The three ecologies are not distinct fields, but areas with undefined borders that are always meant to partially overlap with one another when the critical question of our practices and the forms of awareness that they condition is undertaken.

The urgent ecological transition of our present time demands that we retain this requirement to use concepts that lie outside of the flux of life. We must change our relationship with the world by reassessing the concepts that until now have helped us define what has always posed an impediment to a "good life". This shift involves an existence where our relationship with the world is not determined by an unrestrained desire for domination, where individual

DOI: 10.4324/9781003455523-5

freedom is understood other than as a fatal figure of possession, and where community is the source of sharing, not submission. *Alienation* offers a dynamic formulation of this concept that needs to be addressed to understand what contravenes a good life and to design the tools needed to ensure the ecological transition is more than a simple advertising slogan. But therein lies the risk. The schizophrenia of triumphant capitalism stands in our way. With its ability to turn any authentic need into a commodity, the organic intellectual has only to cover up its deceptions.

A constellation of terms frames this concept to create a genealogy of understanding in which history and anthropology merge. Within this framework, *submission* figures prominently, thus granting a prime position to the question of civil status whose universality is denied under a reign of military conflicts. Alongside it, *servitude* confers a pre-eminent place to work within systems of domination in degrees ranging from slavery to oppression whose characterisation implies a broader subjectivity. As the idea of the salaried worker evolved, new forms of domination appeared, including *exploitation* and *reification*, which give reason the destructive function of reducing each living being to a simple end, even when their position as a legal subject is recognised by all. Finally, there is *indoctrination*, which has acquired an influential role as an instrument for managing the masses over the last century.

Alienation occupies a unique position in this thesaurus of domination. Whilst its origins mean this concept has developed slowly and with difficulty between a legal sector, which alienates property, and a medical sector, where individuals are alienated, both cases draw on the idea of a loss of control over the subject. The alienated subject is an individual who is *a stranger to themselves*. This "foreignness", as Hegel demonstrated, arises from a unique relationship to the world that Bergson, a century later, would term "Homo faber". If the essence of humanity is to become deprived of its essence, to disassociate oneself from the world to build a better world, humankind has the power to bring about its own destruction. Homo faber is not intimately connected to any environment. Technology places Homo faber at the centre and makes life appear as an external element. Homo faber thus created from scratch "nature", to which humans only belong because of a linguistic slippage employed in the phrase "human nature" without considering its aporetic character.

Within the concept of Homo faber, it is not only the notion of work but also a relationship to the world that views all materials, all forms of existence, as potentially usable ends for production. Yet, alienation can be applied to many fields beyond work. Take the example of property where alienation may mean that individuals are unable to distinguish themselves from the material goods they accumulate. From precious metals in antiquity to the most immaterial forms of property today, this mode of alienation reduces the individual to considering their own existence solely by determining a quotient whose logic, whatever they think about it, escapes them. Other examples include the

paralysing dependence on new modes of communication or even a powerless-
ness to resist standardised images of the ideal body. These two contemporary
manifestations of alienation are certainly linked to a representation of work
that pervades contemporary societies and affects the borders between public
places and private spaces. However, they also have a degree of autonomy,
which explains how they penetrate consciousness even where work tends only
to be an illusion. More generally, our world of consumption continues to perpetuate alienation in a paradoxical way: it promotes the uniqueness of the individual to hold them firmly captive to a norm where all people are interchangeable
through their similar desires. Extracting oneself from this norm is painful.
Being an outsider, even when one lives in an isolated area, can make life seem
deprived of meaning.

Before being an operational category for social criticism, alienation was
effectively first tested as part of a criticism of culture understood in the broad-
est sense. Applying the Marxist framework that plays a determining role in this
concept's history, it is clear that understanding the idea of "commodity fetish-
ism", introduced in Marx's *Capital*, relies as much on a rigorous analysis of
economic relationships as on a new manifestation of domination where all
these relationships are simultaneously devoid of sentimentalism and spiritual
references in order to sustain themselves and be rooted sufficiently in culture
for a whole society, including those excluded from it, to devote itself to mer-
cantilism. If capital, beyond exploiting the hard work from which it originates,
acquires a social power described as alienating, it is due to a series of divisions
in the world that attributes value to distinct acts, that distributes their functions, and that hierarchises individuals who find themselves reconstructed as a
unit in a product that thereafter gains a discursive power worthy of writing its
own history.

Homo faber thus possesses the seeds to destroy life. The alienation that conditions Homo faber's relationship to the world can lead to such a degree of
land artificialisation in humans' natural habitats that it can accompany the
worst delusions: a Trump or a Bolsonaro are but two expressions amongst
others. Implementing an ecological transition implies a fundamental questioning of Homo faber's condition, both in terms of our supernatural relationship
to technology and our intimate relationship to the corporeity of a world of
which we are an integral part. This unique foreignness to the world, which
makes us believe that we can possess it at will and transform it by annihilating
entire parts of it, goes hand in hand with a competence that is fragile, evanescent, but also extremely fertile: the imagination. Imagination alone cannot
extract us from what Adorno terms a "damaged life", but it nevertheless constructs the necessary conditions for re-establishing the broken contract with
life. Compared to the other forms of domination mentioned above, alienation
has the particularity of being a collective phenomenon that retains in each of
us a feeling of freedom. In other words, opposing alienation necessitates

constantly uniting collective resistance and individual awareness. It is a form of political and social engagement that can only be undertaken for one's personal tranquillity, that requires sharing common goods, but that cannot come to fruition without a frank discussion about how humans are represented.

It is on this basis that the ecological transition must necessarily be accompanied by an uncompromising criticism of all forms of alienation that, despite ourselves, hold us captive in collusion with a regime that disorientates life until it destroys us. The arts, as they powerfully reveal this form of alienation, are vital, not as a doctrine that would need to be defended, but, in highlighting differences, like Homo faber, that are so engrained in us to the point of their being second nature.

Three Key References

Anders, Günther (2002). *Obsolescence de l'homme*, 2 vols. Paris: Éditions Encyclopédie des nuisances/Éditions Fario.
Gorz, André (1988). *Métamorphoses du travail quête de sens*. Paris: Éditions Galilée.
Sennet, Richard (1998). *The Corrosion of Character, The Personal Consequences of Work in the New Capitalism*. New York: W. W. Norton.

5

ANIMAL

Baptiste Morizot

In strict biological terms, "animal" is the vernacular noun for Metazoa: multicellular, eukaryotic, heterotrophic, and generally mobile organisms. Is there any debate over whether humans are metazoans? It seems not. Yet if one asks the same question of the word "animal", many will respond that humans are not or are no longer animals, even though the term is an exact synonym for Metazoa. Evidently, there are other issues at play besides a problem of taxonomic classification. "Animal" must therefore signify something other than what it suggests. The term has no biological basis: it is an ontological and symbolic construct that transformed the Western world. In traditional Western thought, the animal is a concept that is attributed a negative and inferior value, as the binary opposite to humans in the human–animal couple.

Consequently, the animal as a true biological category, to which bipedal metazoans regularly claim they do not belong, does not exist; it has no real referent. Snow leopards, sea sponges, kestrels, *Canis lupus*, *Homo sapiens* are indeed all heterotrophs. However, no animal exists that would unify them into one coherent category. There are only animals, life forms that are simultaneously related yet incomparable. Our research question thus clearly emerges: if the animal does not exist, why has it always been and continues to be a category on which to base our relationship to the world and ourselves?

A close examination of this issue reveals that there is a name for living beings that do not exist – sphinx, dragon, werewolf, Leviathan, Wendigo – yet play a structuring role in the construction of cultures: chimera. Thus, I would like to suggest that the animal, as a fraught category that simultaneously includes and excludes humans, is a chimera. In the Western intellectual tradition, the human has always been defined in relation to the animal, whether by

DOI: 10.4324/9781003455523-6

opposition, distinction, or affiliation: the animal provides a mirror that reflects the human condition. The fact that a non-existent chimera has become the touchstone of a distinction that defines human nature, our ontological identity, is indeed worthy of investigation.

We could take a lazy modern attitude towards chimeras that stigmatises superstitions and irrational beliefs; alternatively, a more fruitful avenue would be to investigate the origins of chimera. Cryptozoology, or the study of unknown animals, is a scientific investigation of creatures, such as dragons, the Yeti, sphinxes, or the Loch Ness monster, whose existence cannot be conclusively proven. Cryptozoology, a term coined by Scottish biologist Ivan T. Sanderson, attempts to objectively study records of fantastic animal forms called cryptids. "Fantastic" signifies that they do not exist in the same way as, say, a tree sparrow, but they exist in their own way and have existed over time. Though the cryptid does not exist, the idea of it persists. Studying the animal as a cryptid, as the fantastic creature that it is, warrants further investigation to uncover the historic tracks that led the animal to haunt the human past as well as the woods.

To begin this cryptozoological study of tracking down the mythical animal, it should be first noted that "animal" as an ontological category that contains metazoans minus humans does not exist in most hunter-gatherer cosmologies. In other words, it is an ethnographic fact that there is no anthropological invariable that identifies the animal in a hawk or a deer. Amongst Tewa Indian tribes, there is no equivalent word for "animal" (Shepard 2013). In the West, a change occurred at some point. Paul Shepard's hypothesis pinpoints the development of a specific domesticating relationship during the Neolithic period that implies a relationship of domination and heteronomisation with the domesticated animal.

Shepard's thesis at once clarifies the act by which the animal cryptid was invented, and how the other cryptid, the human, was conceived as being on its own path of surpassing and elevating itself above animality:

Before civilization, animals were seen as belonging to their own nation and to be the bearers of messages and gifts of meat from a sacred domain. In the village they became possessions. Yet ancient avatars, they remained fascinating in human eyes. A select and altered little group of animals, filtered through the bottleneck of domestication, came in human experience to represent the whole of animals of value to people. [...] So-called amateurs from the city looked over the farmyard fence at the broken creatures wallowing and rolling in their own filth, and the idea of the bestial animal with unfettered appetites was born. This "natural" state represents the model of the animal in philosophy, what made men understandably yearn for unrestricted freedom.

(Shepard 2013: 228)

Shepard's thesis can be succinctly summed up: representing an anthropological difference, the animal is not an animal itself, but the domesticated animal in a relationship that devalues and degrades it, which was taken as an example and then established as the model for all animality.

The farmyard chicken says to the Neolithic child looking at it, "Your hunter-gatherer ancestors took the wild horse, the crane, the lynx, the beaver, the otter, and the wolf as their models. But you will not become an accomplished being by observing and modelling yourself on the pig in the pen." Sitting Bull and Crazy Horse are individualising names; Voracious Pig, Resting Sheep, and Laying Hen are not. These are the names of de-individualised beings. From the child's point of view, all becomes clear. The pig in the pen is dependant, rendered docile, diminished, and exploited: it cannot be a model of individualisation and accomplishment in and of itself. Phrases like "worth little more than an animal" or "only an animal" are founded on this metaphysical misunderstanding. It is not the case that domestic animals are inferior, rather the relationship in which they are trapped demeans them.

Civilisations that have over-domesticated animals through a degrading relationship, that is those that have transformed and theorised them as "beast" and "bestial", have an urgent philosophical need to radically differentiate between the animal and the human animal. Societies that maintain contact with wild animals or whose domesticating relationships are not demeaning (e.g., the Tuvans of Siberia) speak of their power and grace, of proximity and kinship.

This cryptozoological tracking serves, first, to recognise the weak basis of using the hybrid animal to define who we are as humans through distinction. For two millennia, Western philosophical anthropologies have invariably been defined using this speculative strategy which consists of distinguishing the human from other living beings, even though animals are increasingly being understood as possessing ancient human prerogatives (culture, intelligence, empathy). Twenty-first-century philosophical anthropology would be enriched by taking another approach, by defining humans through their affiliation and constitutive relationship to animals. Defining by distinction seems almost a logical necessity. It is, in fact, a political act. A being's specificity could be just as well determined through the singular way in which it relates to others. What then constitutes a human is their fundamental relationships with non-humans, their affiliations and relationships. Trying to define humans by occult distinction undermines their relational essence.

Second, we should not overburden this small word, derived from the observation of a common ancestry amongst living beings, with hidden metaphysical connotations. Heterotrophs, chordates, and mammals are all animals. We can simply accept that humans are animals. There is no lost dignity or stolen path of ascension in this formulation, nor is it a return to a pure primality. The question is not whether humans are animals like any other, but in which ways they are like other animals. What kind of animality is humanity?

Three Key References

Morizot, Baptiste (2018). *Sur la piste animale*. Arles: Actes Sud.
——— (2016). *Les Diplomates. Cohabiter avec les loups sur une autre carte du vivant*. Marseille: Editions Wildproject.
Shepard, Paul (2013). *Nous n'avons qu'une seule Terre*. Paris: J. Corti.

6

ANTHROPOCENE AND AESTHETICS (THE)

Matthieu Duperrex

The Anthropocene, a notion adopted fairly recently in the humanities and the social sciences, implies an enormous reframing of modernity and its legacy. Chemist Paul Crutzen, a Nobel prize winner for his work on atmospheric chemistry, popularised the term in 2000, firstly in an article in the International Geosphere-Biosphere Programme newsletter and then in a very short yet significant article two years later in *Nature*. Beyond the geological and specifically stratigraphic perspective it offers (Zalasiewicz et al. 2011), beyond documenting major biogeochemical cycles in Earth system science or even ecological changes in biodiversity, soil, and geography studies, the Anthropocene appears as a "bridging concept" (Brondizio et al. 2016) that establishes interdisciplinary connections. Once the Anthropocenic hypothesis is accepted, it becomes extremely difficult to return Earth sciences and social sciences to their respective places. The Anthropocene has thus made a dramatic entrance into contemporary thinking, far beyond the sphere of influence for a merely geological concept.

How does the theory of the Anthropocene relate to the question of aesthetics? There are two issues at stake. First, it is a matter of translating, through the medium of art, the concept of the Anthropocene, its capacity to reframe modernity, its dualisms, its false oppositions, and its destructive effects on contemporary ecological conditions. Second, there is a phenomenological need to acknowledge the phenomena resulting from the world's present ecological situation: global warming; carbon, nitrogen, and phosphate cycles; the sixth extinction. Indeed, the first meaning of "aesthetic" is "to be sensitive to", "to experience". This experience is not easy to comprehend, as we are dealing with hyperobjects (Morton 2018) that confront and exceed our scale of action and perception.

DOI: 10.4324/9781003455523-7

Yet, equally problematic is that the phenomena concerned are not unified, because the Earth is not unified. Saying "protect the planet" represents an abusive globalisation (Larrère 2016) that very often renders us powerless. To understand the scale of experience that is envisaged in the concept of the Anthropocene, we must look to the core activities of the Earth sciences.

Another new problem is this experience is not in the first person, at least not wholly, because "the earthling" is only an actor, a third party to the experience. From an aesthetic point of view, humans can no longer stand in front of an inert backdrop or set, like the *Rückenfigur* in Caspar David Friedrich's famous painting *Der Wanderer über dem Nebelmeer* (*Wanderer above the Sea of Fog*).

On this basis, the aesthetic question changes with the emergence of the Earth as an actor entirely separate from human actions. In essence, the ecological crisis remains unknown if we do not study it through the sciences. Here, the science is understood not as a set of results to be cleansed of empirical residue, but science in the process of investigation and translational research. Works of art, then, cannot be isolated from the question of the Earth sciences.

The Anthropocene's emergence is accompanied by a change in aesthetics that can be characterised in three ways:

- A scientific aesthetics (represented by instruments);
- The capacity for representation in the political sense of human and non-human voices;
- The representation in a more classical sense recognised by the art sciences.

This shift also affects our understanding of what we expect from art and its vocation in the context of new climatic regimes. For the experience of the Anthropocene could be described as a crucial moment, one that both represents a break in intelligibility with the fabric of the world and a developing consistency of an environmental situation where new relationships are to be constructed. Today, what is labelled as "environmental art" or "ecological art", which brings together around 50 years of diverse artistic creation, is to a large extent being reinterpreted through the lens of this fraught existential polarity. Across contemporary art, the recurring motif of an investigation (Duperrex 2018) draws the Anthropocene's aesthetic resolution towards carrying out "art as experience" in the pragmatic sense (Dewey 2014). It is able not only to maintain, but also to invent or reinvent, a place to live within the troubling situation of the new climatic regime, even as a process of experiential knowledge through artistic practice is undertaken.

Finally, if living in the Anthropocene is in some way living in the catastrophe of the present, it supposes a reconsidering of the *ecumene* (Berque 2016) through the triple disenfranchisement – ecological, technological, symbolic – that affects the links between social milieu and environment (Moreau 2017). In that sense, art would employ a residential perspective on the condition of redefining what has been taken away in each of these three dimensions of the human world, hence the importance of a "behavior of art" (Dissanayake 2010). Its collective promise goes far beyond experiencing individual satisfaction by gaining better experience or knowledge of our living environments. This view returns to an affirmation of the artist's responsibility (response-ability) towards their object of study. Beyond aesthetics, the environmental ethics deduced from this configuration of artistic enquiry proceeds not from norms of limiting human action by a transcendental recourse to the global, but from procedures that empirically verify the entangled interrelationships found in the situation experienced. Ethics is therefore a benefit of the complex situation in which we attempt to regulate the pace of our lives through our practices. Ethics is always in discussion and negotiation with the materiality of experience, which is why it is a matter of being able to respond, of a responsibility. Further clarifying the concept of the Anthropocene as it relates to art would therefore guarantee its aesthetic significance as well as its ethical comprehensiveness in the relationship experienced with the milieu.

Three Key References

Nova, Nicolas (ed) (2021). *A Bestiary of the Anthropocene. Hybrid Plants, Animals, Minerals, Fungi, and other Specimens*. Eindhoven: Onomatopee.

Ramade, Bénédicte (2022). *Vers un art anthropocène. L'art écologique américain pour prototype*. Dijon: Les Presses du Réel.

Yusoff, Kathryn (2018). *A Billion Black Anthropocenes or None. Forerunners*. Minneapolis: University of Minnesota Press.

7

ARCHITECTURE (THE AESTHETICS OF ECOLOGICAL TRANSITION IN)

Philippe Chiambaretta

Ecological transition is becoming an urgent need for human beings confronted with mounting environmental crises caused by how they live. Global warming and the reduction of biodiversity are the terrifying symptoms of a rapid destruction of natural environments. This situation, unprecedented in the human age, is encapsulated by the term "Anthropocene".

Since the introduction of agriculture and the domestication of animals during the Neolithic era, human beings have always developed ecosystems to control the biological cycles of domesticated species. For a long time domestication was a balanced and unified relationship between humans and their milieus. Across principally rural environments, development went hand in hand with care; the city remained a minor exception for centuries.

The emergence of a philosophical dualism between nature and culture in sixteenth-century Europe made human beings believe themselves to be the masters and possessors of nature. Humans assumed the singular status of the thinking subject amongst other non-human, living and non-living, objects. Armed with the power of reason, the white man of Western modernity led the colonisation of the rest of the world through this process of domination over the weak. With the advent of capitalism, he drew enormous profits from the unrestricted extraction of natural resources considered to be free and infinite. Due to fossil fuels, industry, transportation, and industrial agriculture, the human footprint would progressively expand and migrate towards an urban lifestyle. At the beginning of the sixteenth century, around 5% of humans lived in a city. This figure was still only around 16% in 1900 before rapidly accelerating in the twentieth century, particularly after the Second World War, to exceed

DOI: 10.4324/9781003455523-8

50% in 2018 (4 billion). Estimates for 2050 approach 70% (around 7 billion).[1] The Anthropocene is a "capitalocene" and an "urbanocene".

A principle of dissociation permeates all human activity: division of areas of knowledge, division of work, division of usages of space. The architectural modernity that triumphed in the twentieth century is a descendant of six-teenth-century philosophical modernity, having adopted its principles and aes-thetics. Architecture freed itself from a rooting in local traditions and sought to be abstract, radical, and international. As the city rose out of a separation of functions, its existence relied on exchange and flux. Yet, the balance of func-tioning in synergy with the surrounding countryside disappeared in the twenti-eth century. The shift towards modernity encouraged a rupture with this ecosystemic vision by separating the functions of the city to technologise its metabolism and optimise its productivity.

Urbanism became functionalist, assigning usages to spaces according to "zoning", which eradicated mixed usages. The Western mode of producing the urban environment developed through a domination based on a technocratic culture of separation: separation of areas of knowledge and expertise (city planners, architects, engineers, developers), separation of power (between pub-lic and private), and separation of actors (investors, promoters, users). It embodies a technical vision that emerges from a deterministic rationale and an exclusively anthropocentric concept of the world.

Architecture seeks to produce beautiful objects, symbols, and isolated ges-tures whose form is a sublimation of the new man, liberated from his natural origins by the talent of a creator. This vision of the architect-artist has led to formalist practices, and unfortunately remains prevalent, particularly as it con-tinues to be taught.

Ongoing climatic changes invite us to act and strive for a relational and inclusive vision of the world, which encourages the development of another paradigm for architecture. The Great Acceleration of human predation, from the second half of the twentieth century onwards, has disrupted the complex dynamic systems governing balance within our shared home, the *oikos* (the root of "ecology"). This calls for a redefinition of all our representations. A new ecology must combine the environmental, social, and aesthetic dimensions and work towards a world that is more durable, desirable, and inclusive for both humans and non-humans, with whom we need to invent ways to cooper-ate. Many civilisations were structured around these ways of thinking before being quashed by the West.

In his recent work on inclusion, the critic Nicolas Bourriaud (2022) reminds us that the West developed an aesthetic principle that illustrates the dualism arising from the separation of nature and culture: the opposition between material and form. The notion of *hyle*, the passive material, receiving *morphē*, an active form, was first formulated by Aristotle and then widely circulated, notably in the aesthetics of Schiller where form represents the spiritual principle that comes from working and arranging the passive material. This duality seems self-evident and encompasses all arts. Bourriaud further explains that this principle can also be understood as the most subtle and pernicious of conditionings, the contribution of art to binary thinking on which the Western and capitalist mechanics of world domination are founded. The material, "nature" demoted to "environment", must submit to the will of the active principle and accept the condition of being the matter on which formatting and specifications are imprinted. The *hyle* and the *morphē* demonstrate that a notion of subjection has been inscribed in the theory of Western art for 2,000 years.

The search for practices and aesthetics that integrate the question of the environment has driven my work as an architect since 2013. This search is also at the centre of the approach taken by Bourriaud, a fellow co-founder of and an important contributor to our biennial publication *Stream*, whose work has enriched my practices since establishing the agency, PCA-STREAM. In particular, the theory of relational aesthetics speaks directly to my intuition that architecture cannot be reduced to a stylistic game, as a facade design competition could be, that it is not a question of building isolated objects but of being connected to the world, of accompanying and responding to its transformations. PCA-STREAM thus rejects formalism and the architect's signature to focus attention on the evolving usages of space and how to create the conditions for these usages by approaching them through collective intelligence.

Discovering a critical apparatus that allows us to reconsider the role of the architect within the world of contemporary art, rather than the fossilised one in architecture, has reinforced my conviction that it was necessary to adopt a pluridisciplinary perspective, both as the cornerstone of *Stream* and the architectural approach, which connects practice and theory, that emerges from it. It was by reading the sociologist Bruno Latour and the ethnologist Philippe Descola, and by speaking with artists like Pierre Huyghe or philosophers like Graham Harman and Timothy Morton, who developed object-oriented ontology, that I became aware that these evolutions in art and thought must be matched with a transformation in architecture. Architecture must participate in an inclusive aesthetics, one which calls for learning to adopt a decentred perspective within a universe that includes non-humans. Based on an expanded vision of anthropologies, such an aesthetics would validate putting an end to the dualisms that structure predatory thinking and would seek their dissolution, thereby freeing ourselves from three centuries of the foundational certainties of Western modernity.

In abandoning the position of a creator in the limelight who produces a beautiful object, the architect must consider phenomena, establish connections between fields of knowledge, and explore the city as a complex, dynamic system, as a living metabolism. Building a city requires a shift from a machinist paradigm to a complex and living one. Advocating for more open and reflexive modes of production is an invitation for actors to step out of their assigned roles. The idea is to migrate from a segmented and top-down linear model towards a bottom-up and circular transversal model. With the unprecedented capacity for measurement that data processing allows, the city-metabolism integrates tracking the flow of mobilities, materials, and energies, as well as a systemic approach that includes living beings and sensory experience. The city-metabolism thus re-establishes a traditional urbanistic approach of layers based on new relationships and connections between them.

Within this framework, the architect becomes the conductor with access to bottom-up collective intelligence, supported by artificial intelligence, which is only an instrument to serve this new ecology. "Less Aesthetics, More Ethics" was the title of the architecture exhibition at the 2000 Venice Biennale. Rather than opposing these two notions, the metabolic approach develops a balance to avoid falling into an aesthetics that is too decoratively mimetic of living beings or into an environmental efficiency approach whose financial and technological dimensions could lead to a new functionalism. Beyond this tension, my conviction is that the true transition resides in a metabolic approach that combines collective intelligence and initiates processes that require us to avoid wanting to frame or dominate the aesthetic result in advance.

Note

1 See, in particular, United Nations, "Revision of World Urbanization Prospects" (2018). Available online: https://population.un.org/wup/

Three Key References

Bourriaud, Nicolas (2022). *Inclusions. Aesthetics of the Capitalocene*. London: Sternberg Press.

Harman, Graham (2018). *Object-Oriented Ontology: A New Theory of Everything*. Toronto: Pelican.

Morton, Timothy (2019). *Being Ecological*. Cambridge, MA: MIT Press.

8

ART AND MILIEU (WORKS OF)

Yann Aucompte

Far from having a programmatic aim, this entry is to be read as a report of observations on the different artistic approaches within the collective "Arts, Écologies, Transitions".

A work can be interpreted in very different ways according to the ideologies and the milieus that frame how it is promoted. After encountering ecological works of art, the following questions came to mind. Which specific method of analysis qualifies this genre? Is it necessary to describe the properties that make a work of art ecological or in transition?

In its historical development, artistic creation has been defined since the eighteenth century through a paralegal metaphor relating to the right to property. In this view, the work constitutes a recognised or established action. It is presented as an act of ideation, materialised by and within an institution: a work of art is any object placed in the world of art. Works in transition are in opposition to the logic of isolating symbolic outputs. In general, ecosophic works in transition do not aim for the autonomy of a work of art; rather they take the situations of intervention as milieus. Since a work constitutes a relation of production, we can speak of it as a collective, societal work that is woven by artists. "Collective" is understood not in the humanist sense, but as defined by Bruno Latour (2018): all beings, human and non-human, animals, concepts, technologies, and physics participate in a collectivity. All beings enact transformative effects upon one another.

From this simple assessment, we can draw two observations relating to the characteristics of works in ecological transition: they constitute an event more than an act and an environment more than an action. The work-becoming-process is constructed over time; it is found in moments-events of reproduction

DOI: 10.4324/9781003455523-9

that no longer propose aesthetic commodified objects. The authorial attitude of the modern avant-garde bends reality at the point where it can distort it, reduce it, weaken it, domesticate it: it emancipates, separates, individualises. In modern and contemporary art, autonomous individuals are fully realised in their detachment and their own uniqueness.

In the ecological work, the fact of working in transition means milieus teeming with individuals who intersect, giving the impression of an overpopulated environment that is territorialised and concrete, attached and dependent. The work no longer puts forward one single point of view; many voices (*voix*) along the way (*voie*) can now be heard. Milieus appear overpopulated with non-human beings who interact with one another and are attached to other beings in an assumed and sometimes very happy dependence. This specific feature of art in transition prepares us for a new diplomacy and a new political configuration. Working in transition shows us that not only do we affect other beings, marking them with our imprint, but that they act upon us as well. In French, the verb *affecter* has two meanings: to touch, as in sensory perception, and to assign a role or a job ("affectation"). Other beings *affect* us in both these meanings of the word. Our relationship to nature and to the social is affected by ecological works. In the transition, all beings are unique, "organiz[ing] their closing (that is to say, their autonomy) in and by their opening" (Morin 2008: 11). All beings are unique because they are the product of interactions. Rather than advocating for the autonomy of a human author, they are the proof of their attachments. All human and non-human beings are subjects of history and arouse our curiosity through their influence on the world. They form an "assembly of the sensory" in our representations. Working in transition unites all beings, unveiling the secrets and innerworkings of their existence, which is polyphonic. In a milieu logic, the power of acting is relative to the quantity of attachments.

An ecosophic epistemology therefore calls for thinking in terms of an ontological plurality (Latour 2013), of realities of relations, of collective action, and of attachments. In methodological terms, this epistemology encompasses the necessity of including numerous institutions: law, art, literature, politics, natural sciences, physics. Each has its own regime of veridiction, its own ontology (Latour 2013). An ecosophic epistemology means these multiple epistemologies are in dialogue with each other: a pluridisciplinary research. Truths must emerge from these encounters, not as a competition to declare the truth but as new weavings of reality. Antagonism and contradictions emerge in the phenomena that we observe. Dialogical thought (Morin 2008: 29–30) considers that, in order to form a coherent and realistic being in scientific descriptions, these contradictions are to remain unresolved. This being emerges through description, from artistic productions to the exiled and diffused authoriality. These descriptions are dependent as much on the artist, institutions, and the technical and physical conditions of possibility, as on audiences, climates, and natural positioning.

In this perspective, it is necessary to consider individuals as natural, human, conceptual, vegetative, fictional, theoretical, ideological, physical, and mental, as much in their own becoming as in their network of influence. It is necessary to describe actors as both capable of actions and the (passive) product of their milieu.

Moreover, a scientific approach must be an aesthetic, implicated intervention, not a distant and neutral gaze, that demands we put into practice an archaeology as well as a plasticity of written forms. These writing processes must make us understand that scientific approaches, in terms of ecology, are not *the* solution, but that, for the most part, they bring about the problems at hand. It is necessary to be explicit about the regime of enunciation adopted, to make it felt in written forms: to produce editorial and written forms that are adapted to our discourse and thus do not confirm the state of power relations, preliminary agreements, and implications that language perpetuates. The normativity of writing and principles of actions that these grammatical and syntactical models oblige us to formulate confirm and reinforce anthropocentric descriptions. We need to radically embrace the fact that we have an influence on the milieus that we observe. Our discourse and our interactions with other fields transform the world. This flaw must become the principal competence of any ecosophic research project, not to provide the means of change but to embody the change itself.

All ecosophic forms of study must subscribe to a research-action approach, conforming to the wishes of Dewey (1980) and Guattari (2018). It must combine practitioners and researchers of different disciplines, politicians, designers, and citizens, as well as a realist philosophy.

Three Key References

Latour, Bruno (2018). "Esquisse d'un Parlement des choses". *Écologie & Politique* 56(1): 47–64.

Morin, Edgar (2008). *On Complexity*. New York: Hampton Press.

Simondon, Gilbert (2005). *L'individuation à la lumière des notions de forme et d'information*. Paris: Million.

9

ART IN COMMON

Estelle Zhong Mengual

Types of art are normally defined by a common medium (e.g., digital art), a common aesthetic style (e.g., minimalism), or even a manifesto to which a group of artists subscribe (e.g., Fluxus). This is not the case with "art in common" (*"art en commun"* in French). To demonstrate, one need only try to define the term using the following list of participatory art projects. In Jeremy Deller's *The Battle of Orgreave* (2001), former Orgreave miners participated in a historical re-enactment of the 1984 confrontation between police and striking workers. In *Enemy Kitchen* (2003), Michael Rakowitz taught his mother's Iraqi recipes to American high school students wearing "Enemy Kitchen" aprons during the Iraq War. In *One Flew over the Void (Bala Perdida)* (2005), Javier Téllez worked with patients from Tijuana's psychiatric hospital to launch a human cannonball over the Mexico–United States border. In *Gramsci Monument* (2013), Thomas Hirschhorn invited the residents of the Forest Houses project in the Bronx to construct a monument in honour of Italian philosopher Antonio Gramsci. In *The Boat Project* (2011–12), Lone Twin built a sailing boat with volunteers using 1,220 wooden objects donated by the participants. In *2 Up 2 Down/Homebaked* (2011–13), Jeanne Van Heeswijk established a community bakery with the residents of Anfield, a deprived suburb of Liverpool. In *Vision Quest: A Ritual for Elephant & Castle* (2011–13), Marcus Coates, with an eagle on his arm, met residents from the Elephant and Castle housing estate in London and carried out on-site shamanic consultations. The extraordinary diversity of media, devices, and contexts is striking. What is the common denominator of these works that justifies my creation of a unique term to describe them? A question. Art in common is a practice that centres on a common question. However, identifying a category of questions, as opposed

DOI: 10.4324/9781003455523-10

to a category of forms, is another way to isolate the heritages and influences within artistic practices. The question is: How can art change life? It is not a new question; indeed, it was at the heart of the avant-garde artistic project. Yet, the response is unprecedented. The best way to understand its specificity is to consider it alongside previous reflections on politically engaged art.

Jacques Rancière views the traditional model of how artistic engagement is received as a mimetic model of criticism. This model operates upon a belief that a work can, if the artist intends it, develop the audience's political awareness. Rancière cites the example of Martha Rosler's *Bringing the War Home* (*c.*1967–72), a photomontage that juxtaposes affluent American homes with images of the Vietnam War. Montage is fertile ground for this type of artistic engagement: the juxtaposition of heterogeneous or even contradictory elements on the same canvas increases "an awareness of the hidden reality and a feeling of guilt about the denied reality" (Rancière 2009: 27). This form of politically engaged art employs a shock strategy, characterised by a critical mode and denunciatory tone. According to Rancière, the limits of this model are found in the absence of a "calculable transmission between artistic shock, intellectual awareness and political mobilization. [...] There is no straight path from the viewing of a spectacle to an understanding of the state of the world, and none from intellectual awareness to political action" (2010: 143). The question that emerges from Rancière's work could be formulated as follows. Is it reasonable to base all politically engaged forms of art, works conscious of producing political effects, on this singular and hypothetical construct of awareness? Is this not a leap of faith? If this is the case, how can we create politically engaged works that do not rely on this mental jump?

Rancière observes that the concept of relational art, as described by Nicolas Bourriaud, attempts to respond to this impasse within traditional politically engaged art. Relational art removes the construct of awareness, the blind spot in the previous model, and proposes that works in and of themselves are forms of relations. It achieves what could only be hoped for once awareness had developed in the first model of engagement. Yet, this model of engagement also presents limitations in terms of its political effectiveness. Relational art, at the same time as proposing forms of relations as works, retains the idea of a work being an object created to be viewed. This notion implies that these relations become visible either when displayed in a museum setting or, if they are intended for public spaces, when they are seen. Choosing to make forms of relations, which then become works, visible is to submit to a regime of visibility. Therein, it fails to create relations effectively and only acts as metaphors or as "the anticipated reality of what it evokes" (Rancière 2009: 77). Rancière continues:

> Art is supposed to "unite" people in the same way [Bai Yiluo] had sewn together the ID pictures that he had previously taken in a photographer's

studio. [...] The concept of metaphor, omnipresent in the rhetoric of the curator, tends to conceptualize the anticipated identity between the form of "being together" offered by the artistic proposition and its embodied reality.

(2009: 77–8)

What makes art in common different can be found at the intersection of these two forms of politically engaged art. Like relational art, community art bypasses the concept of awareness by directly proposing forms of relations as works. Yet participatory art projects are rarely conceptualised as objects made to be viewed: they are made to be made. These works move out of the museum setting, an unordinary space, and are embedded in social spaces, in places of everyday life, which further liberates them from the regime of spectator visibility. Participatory art projects are based on a new apparatus: a co-production or co-creation over time between the artist and participants. Art in common finds a solution to the pitfalls of metaphor that haunt relational art. These works have a smaller risk of anticipating reality as what makes them possible is based on the material organisation of a collective. Art in common thus invents a new form of politically engaged art. Producing political and social effects no longer takes the form of developing awareness amongst viewers or establishing a future collective. Politically engaged art becomes a collective experience *in situ* over time through individualising encounters or new ways of relating to others and to certain problems the world faces. Its power to transform reality is not an effect to hope for, but an effect already in the process of being achieved. Today, through its unique response to the age-old project of changing art and life, art in common reaffirms art as a fertile ground for observing and reinventing possible conditions and forms of collectivity.

Three Key References

Bishop, Claire (2012). *Artificial Hells. Participatory Art and the Politics of Spectatorship.* London: Verso.

Rancière, Jacques (2009). *Dissensus: On Politics and Aesthetics*, trans. Steven Corcoran. London: Continuum.

Zhong Mengual, Estelle (2019). *L'art en commun. Réinventer les formes du collectif en contexte démocratique.* Dijon: Les Presses du Réel

10

BIODIVERSITY

An Aesthetic Emergency

Joanne Clavel

The word "biodiversity" is composed of the Greek word "βίος" (*bios*) meaning life and the Latin word "*diversitas*" meaning variety or difference, as well as divergence. The term defines nature from a scientific perspective within the historical context of a massive erosion of diversity amongst living beings. Biodiversity is not equivalent to "biological diversity", which designates a scientifically objective property of living beings, rather the concept problematises the relationship between that property, which is under threat, and society by implicating the responsibility of humans faced with their own destruction (Maris 2010). Biodiversity thus reveals a concern for the disappearance of other species and associated feelings of dread in the face of human actions. Since its emergence, "biodiversity" has progressively replaced the term "nature" in writing and discourse. Yet, in contrast with the pedagogical, scientific, administrative, and technocratic spheres, the worlds of art have rarely accounted for this linguistic shift. After briefly discussing the term's history and the alarming state of the extinction of biodiversity, I will examine its aesthetic links through the lenses of life forms/forces and experience whilst emphasising the urgent political need for aesthetics to become part of the debates around biodiversity.

In the mid-1980s, a group of biologists in the United States, who were analysing the anthropic impact on living beings, were troubled by the rapid disappearance of their subjects of study in terms of proliferation and population, as well as the complete extinction of some species. For example, the last song of the 'ō'ō'ā'ā (*Moho bracattus*), an endemic bird of the Hawaiian island Kauai, was recorded in 1987 by Cornell University's ornithology laboratory. The male bird sings his melody over and over, but no females come as they have all been

DOI: 10.4324/9781003455523-11

wiped out along with the other members of the Mohoidae family and many native bird species following the islands' colonisation. Outrage at the stories of the depopulation of life on earth that proliferated in research data, archives, and the scientific imagination[1] mobilised scientists to collectively act. A scientific discipline – conservation biology – was created through events including the "National Forum on BioDiversity" (1986) where the term "biodiversity" first emerged (Wilson 1988). Biodiversity, as defined by scientists, is the diversity of all forms of living beings across different organisational levels. Three principles structure the discipline: genetic, specific, and ecosystemic. It also operates on different time scales, including an evolutionary one. Biodiversity is therefore a collection of entities in a dynamic relationship with a past heritage that determines the shape of their future potentiality.

On the scale of life on earth, the polyphony of living beings is disappearing at an unprecedented rate. These thanatopolitics began at the latest during the period of large-scale colonial exploration with its assault on isolated environments and have only accelerated since the Second World War. Scientists refer to the "sixth mass extinction of biodiversity" and believe it to be more intense than any previous ones. The richness of species is threatened – 41% of amphibians, one-third of coral species, one-third of conifers, one-quarter of mammals (IUCN 2020) – and the loss of abundance or defaunation is just as worrying: 60% of wild vertebrates have disappeared in 40 years (World Wildlife Fund 2018). In Europe, a similar collapse is observed amongst farmland bird populations (see Vigie Nature), and even in "protected areas", 75% of the biomass of flying insects has disappeared in less than 30 years (Hallman et al. 2017).[2] Moreover, species that have historically evolved alongside their specific environment are most affected, leading to a homogenisation of living beings (Clavel et al. 2011). Every day, a few more of the earth's ecosystems are destroyed and artificialised for industrial human food production, leading to the current situation where 70% of birds on earth are chickens and 60% of mammals are cattle (Bar-On et al. 2018). If ethical questions have taken precedence in scientific debates as shown by, for example, the debates over the value of diversity at the Intergovernmental Science-Policy Platform on Biodiversity and Ecosystem Services or within the ecologist community,[3] aesthetic questions and in particular our sensorial relationships to living beings are not or rarely voiced in these discussions. Yet, ecologists, as Larrère suggests (Larrère and Larrère 2015; Beau and Larrère 2018), also undertake "an ecology in the first person". For example, we can cite agroforester Aldo Leopold's famous *A Sand County Almanac* (1949), or we can point to how the writer, ecologist, and naturalist Robert Pyle (2003) was one of the first to warn people of how our experiences are impoverished without companions other than humans.

Addressing the aesthetic questions around biodiversity is necessary for reasons relating to environmental justice, survival, and creativity. First and foremost, we must recognise that sources of destruction are not situated in

Anthropos as a uniform species, but in a particular way of acting and doing that is historically, geographically, and socially situated (Bonneuil and Fressoz 2013; Hage 2017). Re-establishing a history of non-destructive practices and lifestyles is necessary to bring justice to those who are the first to be affected by global changes without having participated in them. These historical injustices are found in ways of acting as well as in the economic inequalities within a country or between countries.[4] Moreover, if the concept of nature has been a powerful tool for creating inequality between human beings, that of biodiversity today relies on claims of political autonomy by numerous indigenous communities[5] and can thus form part of a new balancing of local and scientific knowledge in politics. The temporalities of biodiversity and our common filiation as living beings should fundamentally change the Western view of the human's place in the world. The anthropocentred model that situates humans as the only superior beings capable of moral reflection sits in opposition to the scientific data on biodiversity that considers human beings as a species amongst other living beings (0.01% of the earth's biomass) and belonging to the animal kingdom (part of the Hominidae family which, only a short while ago, was itself diverse). This ecocide jeopardises humanity's survival in the short and long term since the human co-constructs itself with and through its environment. The environment should not be seen as simply a backdrop, but rather as a major actor in the evolution of living beings and the personal development of individuals and collectives.

With the agency and opportunities it offers to invent new realms of possibility and new social orders, experiencing biodiversity must be at the heart of environmental creativity. For the slaves who escaped plantations, the tropical forests of the Amazon or São Tomé were places of alliance and refuge where they could create dignified lives (Touam Bona 2021, 2022). It was in the middle of a wood, close to a pond, that Henry David Thoreau, supporter of the abolition of slavery alongside his sisters, developed his political thought of civil disobedience. It was in crossing the forests of the alpine arc where he collected plants that Rousseau conceived of the first political treaty in Europe on social inequalities. It was in the heart of the Cévennes mountains that Fernand Deligny and a small network of colleagues invented new lifestyles and created new ways of establishing communities alongside people with autism or mutism. The experience of living with domesticated biodiversity, such as reinvents itself on the margins of the globalisation of agriculture, will be a contemporary trace of the diversifying and ever-changing sensorial history between biodiversity and humans. Faced with the extinction of biodiversity and in a world where it is necessary to learn "self-defence" (Dorlin 2022), where corporeal energy, movement, and sleep alike are other resources to pillage (Crary 2016), we feel the urgency of an ecological *aesth-ethics* of living beings. The development of such an aesthetics would introduce major transformations in our practices and lifestyles, allowing biodiversity to live on and play its role in the biotic community.

Notes

1 For example, the drastic decrease amongst the halieutic population in marine eco-systems. The baseline for the oceans has not stopped falling and new norms will continue to be set until the day when there will only be farmed fish left (Pauly 1995: 430; Gremillet 2019).
2 In 1962, Rachel Carson warned against the mass destruction of living beings. Even back then, she made the link between an intensive usage of pesticides and the snow-balling extinction of animal populations by removing the insects at the bottom of various food chains.
3 We can see how the controversial "new conservation" demonstrates an official cap-italist utilitarianism.
4 The richest 10% of the world's population (around 630 million people) are respon-sible for 52% of the cumulative CO_2 emissions, which is close to a third of the world-wide carbon budget over the last 25 years alone. The richest 1% of the population are responsible for 15% of overall emissions (see OXFAM 2020).
5 The Convention on Biological Diversity, an international treaty adopted in 1992 at the Earth Summit in Rio de Janeiro, emphasises the positive contribution indige-nous people make to maintaining biodiversity (see United Nations 1992: 6).

Three Key References

Maris, Virginie (2018). *La part sauvage du monde. Penser la nature dans l'Anthropocène.* Paris: Seuil.
Morizot, Baptiste (2016). *Les diplomates: cohabiter avec les loups sur une autre carte du vivant.* Marseille: Editions Wildproject.
Pyle, Robert (2003). "Nature matrix: reconnecting people and nature". *Oryx* 37(2): 206–14.

11

CINEMA

Damien Marguet

Every year, new images of "natural" catastrophes associated with climate change contribute to the iconography of the ecological disaster. Whilst primarily circulating on social media, these representations belong to an imaginary which cinema has helped create. For the writer Richard Flanagan (2020), photos of the huge fires that ravaged Australia at the end of 2019 echoed the visions offer by the *Mad Max* series or Stanley Kramer's *On the Beach* (1959).[1] Yet, as Dork Zabunyan stresses, "rather than provoking a shift in our forms of observation, the hyper-spectacularization of the ecological disaster acclimatises us to it" (Zabunyan 2019).

The disaster iconology corresponds to an aesthetics of the "ecological sublime"[2] from which it draws, by contrast, its power. The juxtaposition of shots celebrating sublime wild nature with another series denouncing the grotesque disaster of humanity in the trailer for Yann Arthus-Bertrand's film *Home* (2009) perfectly illustrates this correlation. That said, how might a relationship between cinema and ecology that would not be classed as "defending nature", whilst still contributing to awakening the viewer's ecological consciousness, be envisaged? This question can be answered by ecology itself, which, from the end of the 1980s onwards, has extended beyond a reflection on the environmental issue. In this regard, Félix Guattari's affirmation of "an ethico-political articulation [...] between the three ecological registers (the environment, social relations and human subjectivity)" (Guattari 2000: 28) announces a shift that seeks to overcome the dialectical ecology of "nature" and "culture". By placing the subject, or more precisely processes of subjectivation, at the centre of the ecological question, Guattari renders it above all a question of *consciousness*. This desire to reinvent our concept of the subject, to affirm its plasticity and adaptability, is found amongst ecology theorists such as Bruno Latour

DOI: 10.4324/9781003455523-12

(2013) and Timothy Morton, who formulated the following equation: "Ecology equals living minus Nature, plus consciousness" (Morton 2010: 19). Access to what he terms "non-identity" passes through a change in perspective and reorientation of the gaze, from blue skies where clear and distinct ideas emerge to the dark earth where all sorts of composite and intermixed species cohabit (see Morton 2009). On that basis, to be ecological means to be attentive to the relationships between entities who are unfamiliar with one another. In these relationships, place and status are unstable, interdependent, and sometimes interchangeable. In the world of images, this implies renouncing any distant or overarching position, such as those that produce the aerial or spatial shots frequently used in environmental documentaries.

Is it possible to identify positions within cinema history and theory that resonate with this redefinition of the ecological field? Animist approaches to moving images, which were very common in writings from the 1920s and have had a veritable resurgence since the mid-1990s (see Sierek 2013), evidence points of contact with this line of thinking. Ascribing the capacity to highlight the personality and energies of non-human entities (animals, machines, vegetation, minerals, and everyday objects) to cinematographic techniques, these approaches view the experience of the film's reception as a chance to transform and increase awareness amongst viewers who discover the phenomena of energetic flow, contamination, and mutation which establish a new relationship with the world. These observations reflect the idea that cinema has a unique relationship to reality. Walter Benjamin (2002c: 115) underlines the fact that shooting a film implies "penetrat[ing] so deeply into reality", an operation generally masked by multiple aesthetic and technical artifices (cutting, editing, retouching, reframing), and it places importance on the significant materiality of the cinematographic image. The camera itself, which makes no distinction between species, reveals the unacknowledged collection of relations and processes (the "tissue"[3] of the world) that we access at the cinema through a two-way movement: the decomposition (cutting) and recomposition (editing) of reality.

In light of these reflections, the camera's ecological properties appear more clearly: it can help to develop new dynamics of perception and subjectivation, and to view and perceive its environment in other ways by discovering strange and unknown elements within the familiar and a part of itself within the foreign. This work implies no longer seeking to mask the cinematographic image's materiality, as the industry so often does, but rather recognising its accidental, plural, partial, and impure character. Moreover, following Benjamin, ecological criticism should focus on all aspects of the cinematographic event and not only the finished work. The working methods of the film director and their team bear witness more profoundly to the ecological dimension of a project than the themes it addresses or the discourse it brings to the screen.

This is precisely the aim of recent research undertaken in the documentary field and presented under the form of an "ecology of documentary methods" (Rasmi 2019). In this approach, the term "documentary" does not refer to a genre, but to a certain way of conceptualising and practicing "creation" that draws on a "loyalty to contingent situations and to presences that inhabit them" and a work of attention "turned towards the exterior and its unstable phenomena" (Rasmi 2019: 18). These features point us to a collection of contemporary cinematic practices that situate themselves precisely *between* documentary and fiction. Whether they are assigned to one of these categories or not, the films of Pedro Costa, Tariq Teguia, Miguel Gomes, Apichatpong Weerasethakul, or even Sharunas Bartas all combine an opening up to contingencies and exercises of attention, and continually maintaining the uncertainty surrounding the spontaneous or constructed, real or fictional nature of their images. The processes on which they draw (long takes and deconstructed narratives, playing with the synchronicity between image and sound, using non-professional actors, territory-based work, and highlighting its folklore) contribute to disrupting the reading of filmed situations and encourage viewers to interrogate their interpretive frameworks.

The film *River Rites* (2011) by American director Ben Russell reflects this approach and includes several aspects already discussed through its references to animism, its appreciation of waiting and chance, and the unique experience of perception that it proposes. Using different formal strategies, the director aims for his work to subvert reading conventions of documentaries and specifically ethnographic film. *River Rites* aligns with this framework since it shows us, through a sequence shot of almost 11 minutes (the duration of a 16 mm reel), filmed with a portable camera, the activities of a group of people on the banks of the Suriname River: several children are bathing and playing by the river, a man is working on a fishing net, women are washing clothes. The perception and understanding of the scene are nevertheless impeded by the director temporally reversing the shot. The logical structure, the relationship between cause and effect in the actions presented, is difficult to grasp. The film confers an essentially mysterious character on everyday actions. Moreover, after a few minutes of strolling along the banks of the river, Russell accompanies these images with noisy rock music whose frantic rhythm seems to match the improbable dives of the bathers. The animist rites of the Saramaka people from Suriname are thus aligned with the trances of American teenagers filmed at a noisy rock concert some years earlier (see Russell 2007). All this contributes to the deterritorialisation of images and of the gaze, with the viewer being led to revise their reading frameworks, their ways of identifying the past and the future, the cultural and the natural, the familiar and the foreign. This work of connection and disconnection through cinema seeks to make our identities more flexible and open up our subjectivities, the only way of extending our ecological consciousness.

Notes

1 Numerous works have studied the evolution of this disaster imaginary in cinema (see Sontag 1965; Szendy 2017).
2 Within North American ecocriticism from the mid-1990s onwards, this concept has attracted particular attention through an examination of the notion of wilderness. See Cronon (1995), which is often considered as the point of departure for this debate, and the edited collection by Callicott and Nelson (1998), which offers a general overview of it.
3 Note that one of the terms that Benjamin uses in German to designate reality "*das Gewebe*", or "tissue" in English, is close to the "mesh" used by Morton (2010: 28) to describe "the interconnectedness of all living and non-living beings".

Three Key References

Cubitt, Sean, Salma Monani, and Stephen Rust (eds) (2013). *Ecocinema Theory and Practice*. New York: Routledge.
Delon, Gaspard, Charlie Hewison, and Aymeric Pantet (eds) (2023). *Écocritiques. Cinéma, Audiovisuel, Arts*. Paris: Hermann.
Narraway, Guinevere, and Anat Pick (eds) (2013). *Screening Nature: Cinema Beyond the Human*. New York: Berghahn Books.

12

CO-CREATION

Collective, Participatory, and Immersive Art

Alice Gervais-Ragu

The artistic field in the last two decades has been marked by collective forms where artists and the public work together to create cultural commons that are both aesthetic and relational. This trend, which we will examine in the field of choreography without being limited to it, has led to the emergence of living milieus where ways of existing, imaginary worlds, and perception systems continually interweave in often complex networks. How these milieus emerge, the perspectives they offer, and the effects they produce will be identified through analysing areas of tension and the critical spaces that they open up.

The economic austerity of recent years and the institutional requirements for artistic works to address the public within a specific framework have generated different interpretations of public participation. For example, some artists target specific groups, like older people with disabilities, and engage them in the process of (re)creating a performance. The dance studies researcher Isabelle Ginot led a university seminar called "Figures faibles" that examined different ways of considering and practising collaborative art forms using the work of three artists who normally employ conventional creative strategies (professional actors, performers facing the audience, theatrical forms).[1] This timely opening up to the participatory field can be observed in the emergence of collectives and artistic approaches that include different contributors throughout, thus bringing together artists, researchers, residents, and specific environments. Amongst them, we can cite the collective "Kom.post",[2] the long-term process-based and family-led work "Rester. Étranger",[3] and the duo "une bonne masse solaire".[4] The latter collaborates with non-human life to forge a relationship with the landscape: their "collaborators" in the form of rocks, rivers, and plants can be affected by meteorological conditions at any given moment. In almost all cases, a new form is emerging where interest lies not in an end, but

DOI: 10.4324/9781003455523-13

in a shared experience that is in the process of being created. These proposals thus throw into question the very notion of a work: whether visual or choreographic, a work as it is commonly understood always produces traces of an object through which the spectator can develop their own means of perception. When a work is created using a participatory process, the notion of a work is inevitably being redefined.

This process-based dimension sometimes conflicts with institutional expectations, such as the requirement to produce a final object or lead an act of mediation. The photographer Sylvain Gouraud's collaborations over the years with different groups (hunters, incarcerated people, farmers) bear witness to this tension between practice and the expectation of production:

> Institutional bodies often envisage art as producing autonomous works, whereas I believe the work is never finished but dependent on multiple views and moments. I don't want to leave an object to speak in my place. I prefer my practices to be produced in the space where they are developed alongside the people involved in them. I often produce a final photographic object to document the process, not as the end work. The gallery then takes on the role of documenting, thus becoming a place of reproduction.[5]

For Marc Lathuillière, another photographer using collaborative approaches, participatory art does not go hand in hand with mediation: the artist is not a social worker who, for example, ensures a collaboration runs smoothly for the whole intervention in the municipality that funds the residency. He asserts that participation is not necessarily consensual and may have manipulative aspects, particularly when there are disputes with the municipal council or relating to the conditions of being hosted. The artist employing participatory methods therefore must accept infringements upon or an expansion of them: the approach is only of interest if the participation happens to blur or displace it. If what we understand by a "work" traditionally relates to the final act, meaning for Lathuillière the object created through a participatory approach, it can also, as Gouraud claims, be located in the process, in the construction of a narrative, the system of exchange established, and the perceptible emergence of shared experiences. These two artists' approaches reveal a tension between participative and deceptive, which lays the ground for a reflection on the limits of findings or what we suppose them to be. The worlds born out of collaborative forms are populated with disagreements and disputes: the approaches only have value because the participatory mode disrupts them.

Considering co-creation also supposes not being misled by the tension between what is important for the artist who develops it and the participants' interests. Anne Kerzerho outlines the requirements for this attunement[6] in an essay about the collaborative framework "*Autour de la table*",[7] which analyses the issues at hand by carrying out interviews prior to the performance: "it is necessary to find an attunement between what is important for the speaker and for the interview's methodology, what it commits to doing. This attunement does not emerge without conflict" (Kerzerho 2020: 54). Even if the initial objectives of a participatory approach are clearly articulated, new issues, including uncomfortable ones, appear frequently in the process: participants deviating from the approach or being manipulated, institutional misrepresentation, and, for forms relating to non-human beings, poor weather conditions and natural phenomena. Nevertheless, from these conflicts, imaginary worlds can be explored, and unique paths reveal themselves. It is in the participants' sometimes-diverging goals that new perception systems are activated. It is through a need to adapt and a reciprocal adaptability that forms of intercorporeality, which, as Christine Roquet proposes, encompass all that happens within the relational field (Roquet 2019), can play out. The uniqueness of these approaches often emerges serendipitously: an artist may have formulated intentions and expectations, but ends up with a form of empathetic, perceptive, and attentional capital that they never expected or even imagined.

In these cases, the notion of a work is necessarily envisaged as a relationship formed with the milieu where it takes on a critical dimension. The most important aspects of the ecological paradigm for artistic practices draw on an implicit relationship and an attunement, which thereafter becomes the work itself as its traditional form disappears. Considering the composition of the perceptive field, a major concern in participatory art, equally supposes a reflection on all that exists between the process and the final object: the experience. The work, displaced through experience, acquires sometimes unacknowledged political connotations: for what it is worth, some shifts are to be expected and it is the experience of living beings which enables that.

Notes

1 Isabelle Ginot is an academic based at Laboratoire Musidanse and the dance department, Université Paris 8. From 2013 to 2015, the seminar "Figures faibles" (meaning vulnerable individuals) was offered to students on the Master's in Dance Studies and the University Diploma in Somatics programmes. It brought together artists, schedulers, and cultural actors around current practices of producing choreographic works that include vulnerable participants (amateurs, children, older people, those with disabilities, etc.) considered as marginal within dominant performance norms. See: http://www.danse.univ-paris8.fr

2 Kom.post, an interdisciplinary collective founded in Berlin in 2009, employs processes of creation that are not orientated towards producing a unique end, rather it continually reconsiders forms and tools in relation to each intervention's context. See: https://kompost.me/

3 "Rester. Étranger", established by Barbara Manzetti in 2014, is at once a home, a family, and a long-term process-based work, which includes migrants, artists, and people passing through, all of whom are co-creators. See: http://rester-etranger.fr/

4 The artistic duo Ambre Lacroix and Kasper Tainturier-Fink, who have backgrounds in visual arts and theatre, formed "une bonne masse solaire". Their thinking and

work are enriched with "images, words, and thoughts; landscapes, collections, and attempts". See: http://unebonnemassesolaire.fr/

5 Unpublished interview with Sylvain Gouraud (by Alice Gervais-Ragu), Paris, September 2021.

6 The notion of affective attunement, theorised by the child psychologist Daniel Stern, postulates that imitation translates the passage of internal states from the mother to the baby and, reciprocally, through a sharing of affects (Stern 1998).

7 "*Autour de la table*", an approach that offers a relational and discursive dramaturgy, was developed in 2008 by Anne Kerzerho and choreographer Loïc Touzé: different speakers, accompanied by an on-site team, create a story from knowledge of movement and the body, which is then shared publicly around a table of six to eight spectators.

Three Key References

Morizot, Baptiste, and Estelle Zhong Mengual (2018). *Esthétique de la rencontre*. Paris: Seuil.

Roquet, Christine (2019). *Vu du geste, Interpréter le mouvement dansé*. Paris: CND.

Stern, Daniel ([1985] 1998). *The Interpersonal World of the Infant: A View from Psychoanalysis and Developmental Psychology*. London: Karnac Books.

13

COLLAPSONAUTS

Yves Citton and Jacopo Rasmi

From Transition to Collapse

We generally speak of "collapse" as designating the sudden collapse across our systems of production, supply, and institutional organisation, a domino effect caused by structural short circuits, such as a financial crisis and/or energy shortages (Servigne and Stevens 2015; Servigne and Gauthier 2018). The principal merit of this perspective on collapse is to cut short dominant eco-negationism by which economic "business-as-usual" dogmas mitigate evidence of destruction in our living milieus.

Four major traits characterise perspectives on collapse in relation to the discourses and agendas that align with ecological transition. First, the *dramatisation* of this transition encourages us to existentially imagine what the sudden and highly likely reversal of our current modes of production, consumption, and destruction in our living milieus would mean. Second, an *urgentism* (both a sense of urgency and an increasing need for emergency measures) is developing in our relationship to the ecological crisis, which, instead of being projected on a far-off and vague horizon for the generations to come, is established as a brutal and imminent collapse within the current decade: the sixth major extinction of biodiversity is already underway. Third, a *practicalisation* of problems prepares humanity to actually inhabit areas that are becoming hostile due to the ongoing destruction. Fourth, there is an increasing *radicalisation* of positions, as most political debates focus on minor issues that lose most of their relevance when judged against the current collapse.

At least two major currents can be identified within the imaginary of collapse. On one hand, individualist survivalism is often associated with right-wing libertarians in North America, who are preparing for the collapse by

DOI: 10.4324/9781003455523-14

stockpiling supplies and munitions in bunkers. On the other hand, ecosophic collapsology, represented by the French Zadist movement, envisages survival as desirable only through mutual aid and commons culture. Collapsonauts align themselves with the second movement, which promotes "living with" the collapse without denying it or being stupefied by it (Semal 2019). The rest of this entry proposes a timely sketch of some artistic practices that align with collapsologist sensibilities.

Musical and Literary Refrains

Following successive waves of apocalyptic fantasies over the centuries and millennia, contemporary collapsology dates back to at least the beginning of the 1970s, with the publication of *The Limits to Growth* report by Meadows et al. (1972). In revisiting musical movements, such as free jazz, punk, post-punk, new wave, indie rock, grunge, noise, or rap, we find the concurrent formulation of slogans (in the lyrics), aesthetic sensibilities (in the choice of sounds), and (anti-)political positions (in the means of production and provocation). These elements prepared the ground for contemporary artivism, that eclectic mix of emerging aesthetics, tactical media, and political intervention. During the last century, artists like the Art Ensemble of Chicago, The Clash, Gang of Four, Talking Heads, Shriekback, The Smiths, REM, Fugazi, Nirvana, Soundgarden, Radiohead, Diabologum, EXPérience, Keiji Haino, Joan of Arc, and Casey introduced and popularised collapsonaut themes and gestures through catchy tunes and choruses (Pierrepont 2021).

Even if most more or less comfortably bent to the commercial dynamics of the capitalist mediarchy upon acquiring an international media celebrity, these artists represent an ideal of operational creative autonomy in a world perceived as already in ruins and spoilt for the future. Beyond any official political positioning on one contemporary media issue or another, they staged and popularised an internal dissidence that challenges the very premises of the capitalist world's modernising order, presenting it as in a state of advanced and irreversible internal destruction.

Today, this attitude – nourished and fuelled by collapsonaut principles and pathways – is found in contemporary French-language literature, with authors like Marcel Moreau, Philippe Curval, Antoine Volodine (with his multiple heteronyms), Alain Damasio, Jean-Marie Gleize, Lyonel Trouillot, Olivia Rosenthal, Nathalie Quintane, Emmanuelle Pireyre, Kossi Efoui, Vincent Message, Les ateliers de l'antémonde collective, or Quentin Leclerc. Across diverse genres (science fiction, objectivist poetry, fictional reportage, survival stories, or mixed media essays), we can trace a common thread, with each text exploring a patch or a particular possibility within a world collapsed due to neoliberalism, colonialism, or anthropocentrism. All these works present possibilities of living with what has collapsed and experiment with diverse ways of collectively navigating it.

Literary fiction – immersed in the concrete and the affective, rather than presiding over underlying concepts, equations, or exponential curves – explores dramatisation and practicalisation. In very different ways, these texts also demonstrate a radicality due to their positioning outside of the normal horizon of beliefs that define a progressive and developmental modernity. Nevertheless, they only rarely exhibit the urgentism typical of collapsologist discourses. This absence invites us to specify a particularity of collapsonaut arts in relation to collapsologist political agendas.

Collapsonaut Arts

Whilst collapsologist discourses, artivist practices, disaster films, and individualistic survivalism are generally situated in a temporality straining under an urgency to act and not perish, perhaps the most interesting artistic works are those that successfully suspend this state of emergency and place us in a temporality, a during or an after, that has already opened up. Considering the work of the musicians and writers mentioned above, the term "collapsonaut arts" would not apply to those that depict a possible transition, the collapse itself, or even the subsequent reconstruction that eventually paves the way to a brighter future. Instead, collapsonaut arts explore, to cite the subtitle of Anna Tsing's (2015) work, the "possibilit[ies] of life in capitalist ruins".

Artists-collapsonauts do not consider the collapse as a terrible moment to endure, but as a milieu, made hostile through the maltreatment it has been subjected to. It is a milieu that we must learn to live with, in which we must create shared duties through highly improbable and always miraculous hacks. Collapsonaut arts are situated alongside practices that already experiment with the disintegration of dominant supply chains, governments, and financing structures, as "Effondrement des Alpes" ("Collapse of the Alps") (2021), a collection of projects between Annecy and Geneva, aimed to do.

Three very different practices can effectively illustrate the concept of collapsonaut arts. First, we can cite a type of contemporary creative documentary cinema, practised by directors like Pietro Marcello and Sara Fgaier with *Bella e perduta* (*Lost and Beautiful*), Pierre Creton with *Va, Toto!* (*Go, Toto*), Nicolas Humbert with *Wild Plants*, and Daniele Incalcaterra and Fausta Quattrini with *El impenetrable* (*Impenetrable*). These directors depict, say, patches of floral life on the outskirts of industrial extractivism sites or striking examples of consumerism in the process of disintegrating as subjects deserving of special attention. Not content with simply documenting stories of alternative lifestyles, these filmmakers often employ modes of production, distribution, and projection that can engage new audiences and new cinematographic economies beyond the industrial circuits, which are ready to collapse alongside everything else (Rasmi 2021).

Second, Dominique Malaquais brought together the artistic practices of a whole new generation of artists and collectives from sub-Saharan Africa in the "Kinshasa Chroniques" exhibition. The exhibition presented examples of performances and works informed by the direct experience of the living-in-collapse that extractivist colonisation has been imposing on the city for decades. The situation is criticised by denouncing developmentism ploys and recycled through reappropriations that transform waste into a weapon of cultural reconquest (Malaquais 2019).

Third, another form of collapsonaut art can be found in Stefano Harney and Fred Moten's *The Undercommons: Black Study and Fugitive Planning* (2013). This essay collection demonstrates how creative escape tactics (or fugitive planning) can cultivate collective thought, poetisation, and improvisation amongst populations sacrificed by modernisation. It addresses not only black victims of slavery in the United States, but also exploited workers in the Third World and all generations to come, regardless of their skin colour. The aim is to escape from the constraints that the worlds of knowledge (the university), art (fairs, galleries, museums), and politics (regulations, media) impose on our collective intelligences and our individual sensibilities (Harney and Moten 2021).

In these three cases, a paradoxical artivism, which simultaneously distrusts the art world and mistrusts political institutionalisation, seeks to put an end to the business-as-usual model by which eco-negationism drip feeds ecocidal regimes of production. With the means at hand, collapsonaut arts are already experimenting with and exploring the bricolage of precarious patches, making local and temporary forms of creative life possible in collapsed environments.

Three Key References

Citton, Yves, and Jacopo Rasmi (2020). *Générations collapsonautes. Perspectives d'effondrement.* Paris: Seuil.

Harney, Stefano, and Fred Moten (2021). *All Incomplete.* Wivenhoe: Minor Composition.

Rasmi, Jacopo (2021). *Le hors-champ est dedans! Michelangelo Frammartino, écologie, cinéma.* Lille: Presses du Septentrion

14

CONTEMPORARY DANCE

Joanne Clavel

How might we conceptualise ecology or ecologies within Western contemporary dance? How can dance respond to the ecocides that the world is facing? Throughout the history of choreography, numerous movements have been described as "natural" or "authentic" due to their awareness of, or challenges to, the acceleration of the modern world, or based on their inclusion of experiences in nature and bodily explorations that embrace the agency of one's own organicity. Today's contemporary dancers present immersions in different environments: site-specific dances (Perrin 2019; Hunter 2017; Barbour 2019) and performances away from the stage, multispecies dance practices, reinventions of body models and ways of thinking about movement, a bricolage of social practices and forms of "artivism". While these experimentations lie on the relative margins of the dominant choreographic forms, they are entangled with other social, aesthetic, and political practices, which makes dance fertile ground for thinking about complex ecological issues.

At the beginning of the twentieth century, Isadora Duncan (1877–1927) described her art as originating "from the great Nature of America, from the Sierra Nevada, from the Pacific Ocean" (Duncan 1998: 223) and a choreography of nature itself springing from "the movements of the clouds in the wind, the swaying trees, the flight of a bird, and the leaves which turn" (Duncan 1998: 153). Her explorations of movement form part of a criticism of practices that constrain the body, specifically those requiring women to wear corsets and *pointe* shoes in ballet. In turn, she proposed a new vision of the dancing body that is fluid and vibrant. During her time in Europe, Duncan read and then became involved with the German biologist and philosopher Ernst Haeckel, considered the father of ecology. Decades later, Anna Halprin (1920–2021)

DOI: 10.4324/9781003455523-15

reinvented the conditions for politically engaged choreographic art. A long-term resident of northern California, Halprin developed an environmentally focussed art form that was sensitive to the relationships between the natural and urban environments and between social relations and current political affairs. To allow more diverse audiences to access her practices, she initiated a bursary system that encouraged social mixing and created the first interracial company in the United States. Halpin worked closely with her husband, architect and land-scaper Lawrence Halprin. From the 1960s onwards, the couple developed a participative and interdisciplinary art form where architects, landscapers, dancers, psychologists, and visual artists came together to create a community and weave art into life. With *Planetary Dance* (1981), which she considers her major work, she contributed to the development of dance rituals that lie at the intersection of care, spirituality, and militancy (Clavel and Noûs 2020). More broadly, the somatic research[1] developed in California (the Esalen Institute or the California Institute of Integral Studies in San Francisco) would influence generations of contemporary dancers, especially those at the forefront of postmodern dance in New York, such as Simone Forti who studied under Halprin for several years.

Whilst Merce Cunningham, Trisha Brown, and even Meredith Monk have been linked to an ecological culture through the way that their work proposes a transformation in individual perspectives from the starting point of holistic or animistic philosophies (Kloetzel 2019), they happily embraced the cultural industry. Nevertheless, postmodern dance experiments with movement, subverting performance codes, and criticises consumer society and North American puritan values. The experimental work of Judson Dance Theater, Grand Union, and Contact Improvisation have recently been analysed as three different forms of anarchist choreopolitics that reinstate the importance of processes that undo hierarchies, institutional links, creative spaces and temporalities, anatomical parts, and relationships to others (Bigé 2020). The dismantling of hierarchies between human and "animal" within an urban context began with the work of Forti who started observing the movements and behaviours of animals in Rome in 1968. There, she amassed the diverse ethnological and kinaesthetic knowledge that would lead to her dance and graphic work. For Forti, "dancing has almost always been a way to explore nature. [...] I identify with what I see, I take on its quality, nature or 'spirit.' It's an animistic process" (Forti 2009: 131). In 1988, Forti moved to Mad Brook Farm, Vermont and worked the land. Her colleagues from Judson Dance Theater, Steve Paxton and Deborah Hay, became her neighbours.

Across time and in different places, movements associated with farming have been incorporated into choreographic scores as a way of creating community and offering alternative economic models to an art market backed by ecocidal capitalism. At Monte Verità (Switzerland, 1900–1920), a small group of musicians, thinkers, and anarchists fleeing the frenzy and pollution of the urban environment engaged in somatic explorations, such as developing new ways of living based on vegetarianism. There, Rudolph Laban developed his theories on movement by reaffirming the importance of agrarian and festival cultures for forms of experimentation and embodiment (Launay 1996). At Black Mountain College (North Carolina, USA, 1933–1957), agricultural practices were also an integral part of the teaching programme and collective life. Yet, neither the community at Monte Verità, nor that at Black Mountain College, formed strong relationships with local farmers, and the autonomous territorial anchorage failed to undo the urban/rural or scholastic/vernacular cultural hierarchies. In the 1980s, Min Tanaka and the Maï Juku dancers established farming practices at Body Weather Farm in Hakushu, Japan.[2] The physical exertions of farmers, their precise and specific movements developed in accordance with each of the plants they grow and care for, became part of the dancers' daily routines. These two art forms that unite living things are gifted practices that can be freely adopted, thereby redefining the contours of a free and autonomous concept of art.

Today's choreography is created in line with the current times and contemporary issues. On stage, environmental themes are becoming more commonly represented through ecological dystopias or the invention of alternative worlds, with the predominant theme being climate change. Yet it is away from the stage that a subsection of the choreographic world has become a laboratory for ecological experimentation. In 2011, Prue Lang, an Australian choreographer based in Europe, staged *Un réseau translucide*, a performance in which the dancers generated physical energy to power the show's electricity. She then proposed a set of green guidelines for her eponymous company. Jennifer Monson created the Interdisciplinary Laboratory for Art Nature and Dance in 2005 after having worked on the navigation systems of migratory animals – whales, pigeons, osprey – (Galeota-Wozny 2006) and created dances that followed their migrational journeys. Other hodological forms have been developed through the act of walking, with pieces by Christine Quoiraud (2000–2007), Agence Touriste (since 2009), and Robin Decourcy as part of *Trek Danse* (since 2010). With or without institutional support, numerous dancer-artists are practising their craft outside, performing their work in gardens, parks, *lavoirs*, and at natural sites (Devigon, Olsen, INUI) or drawing self-portraits of place and new cartographies (Contour, Chariatte). The quality of these enriching movements and practices associated with farming form part of contemporary dance training and inspire creations, as well as addressing demands for decolonial ecology and/or new modes of artistic production (Ferrara, Myrtil, Kerminy art farm). Some are inspired by alterity more than humanity (dance for plants, recoil performance group, Renarhd, or Guérédrat), rural pagan celebrations (Pagès), and diverse human ontologies (Rodrigues).[3] Others criticise inaction against the thanatopolitics of the past and present (Bourges, Rodrigues,[4] Crisp). A plethora of pieces in development are working towards renewing the links between living beings and their environments.

At a time when Western lifestyles rely on a disembodied artificial intelligence that demands our attention and an extractivist regime, aesthetic attempts described as ecological are engaging with environmental ethics, which have the double requirement of embodiment without extraction. Choreographic culture therefore leads to a critique of the productivist model. It is not only about creating beautiful pieces, but also about transforming ways of doing and acting in the world. Drawing on theory-practices of ecosomatics, dance has the potential (as this promise is far from being widespread) to renounce dominant corporal norms. As a pioneer of experimentation in both the relationship between human subjects and nature and the social relations that produce diversity, contemporary dance – as a practice of inventing oneself and the world – can thus be established as a laboratory of emancipation against contemporary biopolitics that destroy the multiplicity and diversity of living beings and homogenise and depopulate the earth.

Notes

1 See Chapter 22 on Ecosomatics.
2 For more background, see Christine Quoiraud's archives at the Centre National de la Danse and the research carried out by her, in collaboration with Alix de Morant, Monique Hunt, and Marina Pirot, for *Le Body Weather Laboratory, Pratique Contemporaine: Un Laboratoire du Toucher* at the Centre National de la Danse in Pantin (2018) and the subsequent research exhibition (February 2020).
3 For Lia Rodrigues, see *Pororoca* (2009), *Piracema* (2011), *Pindorama* (2013), and *Para que o céu nao caia* (2016).
4 See *Fùria* (2018).

Three Key References

Barbour, Karen, Vicky Hunter, and Melanie Kloetzel (eds) (2019). *(Re)Positioning Site Dance: Local Acts, Global Perspectives*. Bristol and Chicago: Intellect Ltd.

Bigé, Romain (2020). "Danser l'Anarchie: théories et pratiques anarchistes dans le Judson Dance Theater, Grand Union et le Contact Improvisation". *Revista Brasileira de Estudos Presença* 10(1). URL: http://dx.doi.org/10.1590/2237-266089064

Galeota-Wozny, Nancy (2006). "Bird Brain Dance: entretien avec Jennifer Monson". *Nouvelles de Danse* 53: 212–36.

15

DECOLONIALITY

Nathalie Coutelet

In 2016, the artist Kader Attia founded ~~La Colonie~~, a space for contemporary artistic practice, panel discussions, and conferences. The written form of the venue's name, in line with its underlying aims, illustrates the persistence of colonial relationships of power, even though many countries gained independence decades ago and territorial decolonisation appears to have been achieved. Kader Attia examined colonial issues and their current hegemony in installations such as *Demo(n)cracy* (2009) or *Les Entrelacs de l'objet* (*Interlacing Objects*) (2020). In his latest video installation, exhibited at the Kunsthaus de Zürich, Attia interrogates the recovery of works stolen from colonised people by drawing on the perspectives of historians, philosophers, activists, economists, and psychoanalysts. ~~La Colonie~~'s other activities have included the decolonial school "Décoloniser l'anti-impérialisme" ("Decolonising Anti-imperialism") (1 December 2019) and "Langage et racialisation" ("Language and Radicalisation") (7 March 2020), a training workshop on the process of "creolisation" and "whitisation" of language run by the organisation, Université buissonnière.

Decoloniality is often associated, or indeed confused, with postcolonial studies, which identifies colonial mechanisms and legacies in former colonies. Established in the 1990s, decolonial approaches emerged from the multidisciplinary work of the Latin American group Modernidad/Colonianidad (Modernity/Coloniality). Their methods reveal the persistence of racial, economic, and cultural domination. Decoloniality's intersectional approach decodes oppression associated with race, class, and gender. The prefix "de" alone speaks to the project of deconstructing and overcoming the colonialism that, even long after the end of colonial empires, is still at work in language, thought processes, economics, and politics. In essence, decolonial

DOI: 10.4324/9781003455523-16

thought does not consider independence to be the end of the colonial system. On the contrary, the system of coloniality persists, especially through political and cultural hegemony. Decolonial approaches therefore facilitate the existence and visibility of other types of knowledge, the production of new ways of understanding the world, and the deconstruction of a hierarchy of knowledge forms developed under the aegis of coloniality. It criticises the universal aims of former colonial empires in favour of highlighting and giving voice to new epistemic productions by descendants of colonial countries or MANA, an acronym proposed by Gerty Dambury in *Décolonisons les arts!* (*Decolonise the Arts*) to refer to individuals from Maghrebian, Asian, Black, or other minority backgrounds.

The initial development of decolonial approaches by intellectuals like Anibal Quijano, Enrique Dussel, and Walter Mignolo associates modernity with coloniality and proposes an alternative to the predominant Eurocentric vision. Ngũgĩ wa Thiong'o, following Frantz Fanon's *Black Skin, White Masks*, seeks to decolonise the mind, to cite the title of his 1986 work, through deconstructing neocolonial political, social, and cultural mechanisms. In 2015, the collective Décoloniser les arts put forward a similar proposal. Author and theatre director Éva Doumbia, a member of the collective, presented Afropean[1] artists as part of the festivals "Africa Paris" (at Carreau du Temple, 2015) and "Massilia Afropea" in Marseille (at Friche Belle de Mai, 2016 and 2018). Her company brings the works of contemporary black authors, such as Maryse Condé (*Ségou*) or Fabienne Kanor (*La Grande Chambre 31*) to the stage. In these theatrical adaptations, the history of slavery and colonialism in France is interwoven with the everyday situations experienced by people of colour in order to highlight the processes of domination and invisibility.

Decoloniality is a much more controversial concept than postcolonialism. Undoubtedly, this is due to its activist and militant aspects, but it also encounters resistance for throwing into question the West's dominating position. In November 2018, a group of intellectuals used an opinion piece[2] to denounce decoloniality as a violation of the principles and values of the French Republic, particularly universalism. Decolonial approaches highlight the concept of pluriversalism, which offers a way to overcome the paradox of universalism by incorporating the reality of the world's plurality, as well as the impasse of binary oppositions, such as North/South, East/West.

From its very beginnings, decoloniality has been expressed through the arts in particular. In 2010, Walter Mignolo and Pedro Pablo Gómez opened an exhibition on decolonial aesthetics ("Estéticas Decoloniales", 2012) at the Bogotá Museum of Modern Art with a workshop led by the curator Elvira Ardila. In his academic work, Gómez is particularly interested in the decolonisation of aesthetic theory and decolonial re-readings whilst also considering practices of resistance to coloniality in all their diversity. In France, the multilingual journal

Minorit'Art, subtitled "a journal of decolonial research", aims to fight against "the coloniality of the emotions" and promote the visibility of "dominated aesthetics". The cover of the second issue "Colonialité esthétique et art contemporain" ("Aesthetic Coloniality and Contemporary Art") features Jean-François Boclé's work *Tears of Bananaman* (2017). The installation creates a human body from 300 kilogrammes of bananas inscribed with messages such as "A Fruit, a Gun" or "Caniba, Cariba, Caraïbe". The banana, as an emblem of colonial construction and its continued endurance as well as colonial exploitation more generally, writes global capitalist markets into the history of slavery.

Bananas, alongside other foods, appear in *Autophagies* (2017) by Éva Doumbia and Armand Gauz. This on-stage cooking performance allows the audience to experience coloniality through their senses and presents the highly political dimension of our meals. Food serves to raise questions about the climatic and human consequences of production. The maafe, shared at the end of the performance as a culinary eucharist, retraces the historic and contemporary journey of the peanut and its imposition as a monoculture on sub-Saharan Africa to meet France's oil needs after 1945. The eucharist points to the often-neglected aspects of decoloniality: bringing people together in a conscious and informed way; going beyond a critical cosmopolitanism of dichotomies; a "transmodernity" (Dussel 2009: 111–27).

Notes

1 The adjective Afropean stresses shared African and European origins. Léonora Miano uses the term in her short story collection *Afropean soul et autres Nouvelles* (2008) and in her essay *Afropea: Utopie post-coloniale et post-raciste* (2020).
2 The opinion piece, published in *Le Point* on 28 November 2018, was signed by, among others, Elisabeth Badinter, Alain Finkielkraut, Gaston Kelman, Pierre Nora, and Mona Ozouf.

Three Key References

Cukierman, Leïla, Gerty Dambury, and Françoise Vergès (eds) (2018). *Décolonisons les arts!*. Paris: L'Arche.
Dussel, Enrique (2009). "Pour un Dialogue mondial entre traditions philosophiques". *Cahiers des Amériques latines* 62: 111–27. URL: https://journals.openedition.org/cal/1619?lang=en
Thiong'o, Ngũgĩ wa (1986). *Decolonising the Mind: the Politics of Language in African Literature*. London: J. Currey.

16

DEGROWTH

Kostas Paparrigopoulos and Makis Solomos

Degrowth

In the 1970s, the mathematician and unconventional economist Nicholas Georgescu-Roegen (2011: 91) wrote that "we should cure ourselves of [...] 'the circumdrome of the shaving machine', which is to shave oneself faster so as to have more time to work on a machine that shaves still faster, and so on *ad infinitum*". Degrowth is not an abstract notion. It makes sense. Only orthodox economists consider it an aberration and wish to convince us of the opposite, that growth is natural. The term does raise some eyebrows, but nothing is stopping us from using agrowth, non-growth, growth objection, or post-growth instead.

"Less is best" is degrowth's guiding principle. However, it does not equate to a simple reduction; rather it is a qualitative transformation:

> The objective is not to make an elephant leaner, but to turn an elephant into a snail. In a degrowth society everything will be different: different activities, different forms and uses of energy, different relations, different gender roles, different allocations of time between paid and non-paid work, different relations with the non-human world.
>
> *(D'Alisa et al. 2015: 30)*

In a degrowth society, several sectors, such as art, education, and health, will "flourish", rather than "develop". Degrowth in the sense of *Homo economicus* is to flourish qualitatively. Serge Latouche (2006: 16) proposes "decolonising the imaginary" to withdraw from the false rationality of *Homo economicus* and "economic totalitarianism". Degrowth theory abandons indicators like GDP, which is "indispensable for managing the monetary economy, but has no

DOI: 10.4324/9781003455523-17

meaning on a political level", so "it wrongly makes us assimilate an increase in goods with the quality of life" (Roustang 2006: 144), to appeal to another economical model: an economy of solidarity and commons, an economy not based on goods. Such an economy "is not an ideal economy. This economy exists as much on a theoretical as a practical level, but it is hidden from the dominant versions of economy provided by most media as well as within research and in universities" (Laville and Cattani 2006: 22). Zones to Defend, the most well-known of which is Notre-Dame-Des-Landes (see Mauvaise Troupe 2014 and 2017), are organised on this economical model. These "real utopias" to use Erik Olin Wright's phrase (2010) develop from other values, such as care and autonomy: "we can conceive of the path leading to degrowth as a path aimed at restoring autonomy as well as a process of liberation from dependence on alienating and heteronomous systems" (Deriu 2015: 57).

Degrowth in the Arts and Music

In the visual arts, questions of degrowth have already emerged with clarity. Turning their backs not only on the circus of the art market but also on institutional art, artists are reconnecting with art as self-production. In parallel, numerous artists have been campaigning in favour of degrowth, without completely renouncing the art fair circuit, since the emergence of the Arte Povera movement at least. The features that characterise these practices include: a shift towards scaling down and simplifying the image as well as more subtle works; working with the territory by using living nature as material; recycling and condemning waste; a preference for collective creations that transform and surpass individual approaches; an opposition to the workings of the art system (see Barbanti et al. 2012: 123).

In music and sound art, questions of degrowth have been undertaken less frequently. Degrowth is specifically concerned with the technology on which music is dependent as much in terms of its production – wave field synthesis, for example, is a very costly sound spatialisation technique – as for reproducing and listening to it. Therein lies the complexity of the issue. In general, music has always been considered a very "technical" art. Yet, several musicians are striving to limit their usage of technology, sometimes lo-fi or self-produced, by using less onerous methods that allow them to recover a real sense of autonomy: an autonomy of production methods. The Italian electronic composer and performer Agostino Di Scipio employs complex technological devices but makes the best use of these means. Moreover, Di Scipio always indicates on his scores how others can reconstruct his systems using open-source software, such as Pure Data, based on non-proprietary and digitally transparent development models. Another example is the French composer Jean-Luc Hervé whose *Germination* (2013), a piece composed for the elite music technology institute Ircam, invited the audience to finish the concert on Place Igor-Stravinsky with a sound system integrated into vegetation. The sound system was made up of small, low-cost MP3 players, each linked to mini-speakers, which gave the impression that a multitude of insects were hidden amongst the plants at the audience's feet.

Do It Yourself, Hacking, and Real Utopias

Amongst the other developments in music and sound art where degrowth trends might be found, we can highlight artists working within Do It Yourself (DIY), hacker, and/or maker cultures, even if they are not (yet) using the term "degrowth". This lineage stretches back to Luigi Russolo who created his own musical instruments, the famous Intonarumori. Pioneers of these practices in the United States include John Cage, the inventor of the "prepared piano" in 1938, and musicians like Alvin Lucier and David Tudor in the 1960s and 1970s. David Tudor's *Rainforest* (1968–73) is an emblematic work within the history of DIY in music and sound art. Punk subculture is also strongly associated with DIY.

More recently, Nicolas Collins's book *Handmade Electronic Music: The Art of Hardware Hacking* (2006) has become a major reference work in this area. The book describes how to make and/or reroute different electronic circuits quickly and with few means to create music: contact microphones, electronic oscillators, mixing tables. Its third edition (2020) was significantly revised and expanded with contributions on new and historical projects by musicians from the international community. Indeed, a significant number of musicians or sound artists are practising the arts of DIY and hacking. For example, John Richards founded the Dirty Electronics Ensemble, a group that emerged from a workshop in which each participant made their own instrument. The idea was to push music into the realm of shared experience, ritual, action, tactility, and social interaction (see Richards 2017).

These hackers, makers, and, more generally, DIY enthusiasts may dream of another world, which is what aligns some of them with those who subscribe to degrowth practices. They strive to recycle and create social and cultural commons. Citing Michel Lallement's characterisation of hackerspaces (2015: 414), we can draw the hypothesis that "the ethics that inspire their members and the creations they champion indicate to us that the time of utopias has undoubtedly returned".

Three Key References

Collins, Nicolas ([2006] 2020). *Handmade Electronic Music. The Art of Hardware Hacking*, third edition. New York: Routledge.

D'Alisa, Giacomo, Federico Demaria, and Giorgos Kallis (eds) (2015). *Degrowth. A Vocabulary for a New Era*. London: Routledge.

Pardo, Carmen (ed) (2016). *Art i decreixement/Arte y decrecimiento/Art et décroissance*. Girone: Documenta Universitaria.

17

DIGITAL CREATION

Anne-Laure George-Molland and Jean-François Jégo

For artistic approaches that are developed and enriched through contact with digital tools and technological innovations (computer graphics, interactive devices, hypermedia, and networked art), directly addressing the environmental question is the equivalent of sawing off the branch one is sitting on. In a frantic race for innovation and increased performance spurred on by the logic of economic growth, technological advances contribute to an unsustainable exploitation of natural resources.

The disciplines in which the arts emerge, particularly digital art, must confront the discomfort and inherent paradoxes of the carbon footprint of the tools on which they rely and the downward spiral of industrialising cultural production by feeding mass-media consumption. Quantitative studies carried out by the think tank Shift Project reveal alarming data relating to the environmental impact of digital technology (Ferreboeuf 2018; Ferreboeuf et al. 2021). In 2019, the production and use of digital systems, linked to the proliferation of peripheral hardware and the explosion of data traffic, accounted for 3.5% of worldwide carbon emissions. At the same time, systems for recycling electronic waste are unfit for purpose (Bihouix 2014).

Against cornucopian discourse, adopting an approach of digital restraint – or an even more radical position – offers hope that we can avoid the catastrophic scenario of doubling the digital carbon footprint by 2025.

Rethinking Practices

Raising awareness of ecological issues must lead to each of us rethinking our practices. Such a step requires time to absorb the knowledge necessary to fully undertake this question, but costed and accessible comprehensive overviews

DOI: 10.4324/9781003455523-18

are now readily available (reports from Shift Project, books by greenIT or ADEME, etc.). Without waiting for institutions to act, individuals are coming together to further developments in their respective sectors. In the academic world, the collective of researchers "Labos 1point5"[1] lists existing initiatives and invites challenges to the meanings given to research. It is our responsibility to reconsider our disciplines and further our teaching practices in accordance with students' needs.[2]

Digital art practices can promote the implementation of innovations, yet their relevance must be challenged on the grounds of environmental cost. The perspectives opened up by 5G, in terms of streaming and real-time media, video games, and soon virtual reality experiences, aptly illustrate the issue of interdependence between technological advances and artistic production. Art and technology reciprocally fuel each other, and artists within cultural industries contribute directly to the increasing production of content and indirectly to the technology race. Certain artists are trying to raise awareness of the ecological cost of different components, software, and materials, as well as their social and ethical costs. Kate Crawford and Vladan Joler's map *Anatomy of an AI System* (2018) dissects the unacknowledged resources involved in making and using an Amazon Echo smart speaker and the utopian dream of being able to recycle such objects.

Beyond the discourse of raising awareness, is it not also time to profoundly reconsider artistic practices by acknowledging the world's energy constraints? Processes that create something new from constraints, like John Cage's *4'33''* (4 minutes and 33 seconds of "silence"), are a source of inspiration. Many artists have shown that a constraint, far from being an imposed stranglehold, can have the opposite effect and inspire inventiveness, as the classical example *Ouvroir de Littérature Potentielle* (Oulipo) illustrates. In a context where the greenest computer is the one that is not produced, the idea of no longer using certain tools or imposing environmental constraints can prefigure a post-digital current, which can be approached from different angles.

Cultivating Frugality

Artistic practices can use alternative processes to the dominant model of ever-increasing innovation to limit their carbon footprint. For example, the principle of frugal innovation invites us to rethink the complexity and cost of processes of creation/production, diffusion, and art markets in environments with limited means. Combining at once technical skill, economy of means, and a natural and intuitive approach (Qadir et al. 2016) reflects ingenuousness, observed in countries in the Global South and in third spaces of creation, like fab labs or makerspaces. Based on principles of inclusivity that bring together actors from all walks of life, this approach favours the capacity to do it yourself. Mirroring practices from DIY communities, frugal innovations draw on notions of autonomy, made-to-measure solutions, sustainability, and common goods.

A recent example is Jerry Do-It-Together (Petit 2015), a computer assembled by groups in several African countries from a plastic container and components of discarded electronic equipment. This project is a source of inspiration for giving a second life to these materials whilst making ways of communicating and digital creation accessible to the greatest number of people. The production of these computers in collective workshops includes customisation to add a unique touch and an intrinsic affective value to a technical tool. Above all, this is a reminder that, before creating, the artist can make their own tools, become attached to them, and therein increase their sustainability.

Making a Material Impact

Reconsidering digital artistic practices also necessitates making dematerialisation tangible. Recent works of digital art have sought to raise public awareness of the carbon impact of data streams, like the collective LarbitsSisters's initiative *BitSoil POPup Tax & Hack Campaign* (2018). This installation materialises the energy used during an interaction with participants by emitting noise, smoke, and continually printing the information exchanged on paper. Whilst rematerialising creative processes comes back to raising awareness of their environmental footprint, artists trained in this area should also regularly undertake energy audits in order to ultimately revise their working practices. Some professional sectors provide the tools to make such calculations, like the application Carbon'clap from the Ecoprod[3] collective, which supports rethinking and optimising filming practices.

Making practices tangible also invites us to reinvent our vocabulary so as to acknowledge the materiality of our resources and supports. Is it not time to explore alternatives to using metaphors in computing? The evanescent image of "cloud computing" could be replaced with terms that explicitly state the impact of its essential infrastructure. Following the artist Julien Prévieux, whose creations raise awareness through the absurd or the overuse of new technologies (Prévieux 2017), cloud computing could be renamed "datacentre computing" or even "global warming computing".

Re-naturalising Time

Rethinking digital artistic practices also means reconsidering our relationship to time. As early as 2007, Edmond Couchot warned of the trend for an ever-increasing functioning in real time: "all our activities, from politics to the economy via the most banal daily elements of work, data, hobbies, or culture, are in a permanent and feverous impatience that tolerates no mediation, no delays in the exchanges" (Couchot 2007: 8).

The current appetite for developing real-time computing to make animated films is an illustrative example of this dynamic. This major optimisation technique raises the political question of how this machine-time should be reappropriated within an economic context that perpetuates the idea of growth. The reduction in time spent generating digital images, a gain in efficiency synonymous with saving material resources, will likely lead to reducing production time and project costs. The time gained will be immediately reinvested in other projects to maintain the studio's productivity levels. Such choices are founded on maintaining the "technological compression of time" (Couchot 2007: 8), resulting in a frenetic race for the moment after, the next production, and eclipsing an appreciation of the present and the satisfaction of a job well done. In contrast, taking an ecosophic perspective, the machine-time gained would not change the total time dedicated to a project. Instead, it would be returned to the artists in a bid to improve the conditions of creation and respect artistic intention.

Fundamentally, it is necessary for all of us to work together to renaturalise time, currently dictated by synchronised machines on a global scale. Reconsidering time means committing to recovering a biological rhythm even more in tune with our normal behaviours, including doubts and hesitations, thereby privileging serendipity and all the detours that are necessary and useful for artistic creation.

Notes

1 See "Labos 1point5": https://labos1point5.org/ (accessed 11 January 2021).
2 See, for instance, the international movement "School Strike for Climate", also known as "Fridays for Future", started by Greta Thunberg in 2018.
3 See http://www.carbonclap.ecoprod.com (accessed 11 January 2021).

Three Key References

Bihouix, Philippe (2014). *L'Âge des low tech: Vers une civilisation techniquement soutenable*. Paris: Seuil.

Ferreboeuf, Hugues, Maxime Efoui-Hess, and Xavier Verne (2021). "Impact environnemental du numérique: tendances à 5 ans et gouvernance de la 5G". URL: https://theshiftproject.org/wp-content/uploads/2021/03/Note-danalyse_Numerique-et-5G_30-mars-2021.pdf (accessed 27 January 2023).

Radjou, Navi, and Jaideep Prabhu (2015). *Frugal Innovation: How to do better with less*. India: Hachette

18

DOCUMENTARY ARTS

Soko Phay

The tenth and eleventh editions of the "documenta" contemporary arts exhibition, held in 1997 and 2002 respectively, established an era of the document in art. Documents no longer simply provide a trace of reality, but are used in and of themselves as media and material, thereby calling into question the traditional definition of the documentary arts as non-fictional representation. This "documentary turn" indicates a change in practices: it emphasises the transformation of usages, the passage from one art to another, and, in sum, the work's hybridity. Against the theory of art for art's sake, current documentary practice is transartistic; it takes on multiple forms. Its strength comes from this heterogeneity.

In the Face of Extreme Violence

Even if documentary arts have a relationship with reality, which, for Tzvetan Todorov, implies a "referentiality pact" (Todorov 1982: 7), this is no longer a guarantee of fact. Documents as evidence are reconsidered, distorted, transformed. Artists are using documents less for their informative value and more for their ability to construct reality, to open up other imaginaries. In this entry, I will focus on how documents create art, specifically when they are evidence of extreme violence. When faced with the erasure of bodies and any other traces, how do artists appropriate documents to make sense of reality?

Three methods of working that bring together politics and ethics can be identified: approaching the documentary arts through the lens of the witness (employed by Rithy Panh), the critic (employed by Haroun Farocki), or as a fable (employed by Walid Raad). In the interests of accuracy, the first draws on witness testimony to reconstruct the reality of genocide. The second, that of

DOI: 10.4324/9781003455523-19

the critic, distinguishes itself from the first by calling into question official representations and dominant discourses. Third, in the fable, the document becomes a producer of fiction by upholding a power of indeterminacy between the factual and the fictional.

The Witness

When the documentary arts create works out of testimony, the referentiality pact contributes to a regime of veracity. This remains the case even in *The Missing Picture* (2013) where Rithy Panh creates dioramas and clay figures to compensate for the lack of archives relating to the Cambodian genocide. Renouncing a search for the missing images that would provide evidence of the crimes committed, he chose instead to combine his personal testimony and a montage of heterogeneous documents: photographs, televised news programmes, fictional works, and Khmer Rouge films. The inlay of archival extracts into reconstructed scenes and the animation of figurines resembles marquetry in its placing of elements of different origins.

From this fusion of materials, Rithy Panh creates an "archive-work", which does not replace the traces and objects that have disappeared; rather, it makes the viewer a witness to events that are not inscribed in the official version of history. "Archive-works" make the absence and inaccessibility of the past ever more present because they attempt to compensate for this immediately acknowledged lack through creation and storytelling. By introducing fiction at the heart of *The Missing Picture*, the director can draw out the opposition between a traditional documentary approach, based on the empirical reality of the world, and an artistic approach, based on the processes of turning history into story.

The Critic

In contrast to Panh, who constructs a film as testimony, Farocki favours a critical angle. Farocki's aim is not only to question the reality behind images of the past, but also to interrogate how that past is constructed using the very same images. This can be seen in *Respite* (2007), which brings new life to sequences filmed in 1944 by Rudolf Breslauer at the Westerbork transit camp in the Netherlands. Breslauer, a Jewish photographer and prisoner, whom the viewer later learns endured a tragic death at Auschwitz, was ordered by the camp's SS commander Albert Gemmeker to film everyday life for those interned there.

Farocki's aesthetic decisions are radical: showing, for example, the rushes in their entirety without cuts or additions in the first part of the documentary. With no voiceover or music to accompany these raw images, the audience are free to construct their own views. Only the intertitle cards provide complementary information. This contrasts with the second part where the director shows the same images again, but this time his analysis discloses their conditions of readability. The critical dimension comes from how he plays with repetition, slow motion, and close-ups. The name found on a deportee's suitcase, for example, reveals the precise date that a convoy was filmed: 19 May 1944. Another method is playing with associations: he places in parallel a sequence showing the prisoners at work recycling cables and cardboard boxes of "the hair of the living and the bones of the dead" at Auschwitz. *Respite* unearths a repressed past by showing the history that took place off-camera: there is a sense of *solidarity* between the silent images of Westerbork and the missing images of the death camps where most of the prisoners would eventually be sent.

The Fable

Discourses that accompany the traces of and documents from the past show how far committed crimes escape understanding. No consolation is possible, especially when there is a state-imposed amnesia as is the case in Lebanon. Due to the adoption of an amnesty in 1991, the topic of the civil war (1975–91) remains unaddressed. The quest for meaning becomes problematic when the process of mourning is inaccessible due to denial and the impunity enjoyed by those who perpetrated collective crimes. Faced with this state injunction of "not to forget to forget" (Ricœur 2001: 454), Walid Raad's The Atlas Group Archive (1989–2004) produced fictional archives by borrowing historical facts that, rather than attesting to their own truth, counter the fantasy written by state powers in the name of national reconciliation.

Documents discovered, distorted, or devised function as instruments of doubt. The strategy adopted of erasing the border between document and fiction allows for a glimpse of other perspectives that shed light on Lebanon's complex history. Within this framework, *Secrets in the open sea* is an emblematic work. According to the text published on The Atlas Group Archives' website, the collection consists of 29 monochrome blue photographs that were discovered under the rubble of war-ravaged Beirut. When the group sent six prints for laboratory analysis, small black-and-white portraits emerged. All of the men and women pictured were identified as individuals found dead in the Mediterranean between 1975 and 1991. The blue of the seabed acquires the status of a "document" in becoming the repository of a latent image and a history.

Raad's work on city life after the coup of Beirut reveals what is latent in collective history: time is suspended; the capital is plunged into a coma. This "collective latency", a concept introduced by psychoanalyst Eva Weil (Weil 2000), visible in *Secrets in the Open Sea*, is a call to the silenced voices of the disappeared.

Whilst their aesthetics may differ, Panh, Farocki, and Raad invent documentary methods according to the modalities of the witness, the critic, and the fable, all of which expand the horizon of reality beyond straight factuality. The three directors show that, when faced with extreme violence, the documentary arts are not only concerned with transcribing reality, but also reconfiguring, in the sense of narrative invention, the possibilities of history. Documents, whether reused or reassembled, are shifters of thought, developing meaning precisely where there has been an act of annihilation.

Three Key References

Caillet, Aline, and Frédéric Pouillaude (eds) (2017). *Un art documentaire. Enjeux esthé-tiques, politiques et éthiques*. Rennes: Presses Universitaires de Rennes.
Caillet, Aline (2014). *Dispositifs critiques. Le documentaire, du cinéma aux arts visuels*. Rennes: Presses Universitaires de Rennes.
Chevrier, Jean-François, and Philippe Roussin (eds) (2001). "Le parti pris du document: littérature, photographie, cinéma et architecture au XXe siècle". *Communications*, 71 and 79.

19

ECOCRITICISM AND ECOCINEMA

Cécile Sorin

"Ecocriticism", a term that first appeared in the 1970s, addresses "the relationship between literature and the physical environment" (Glotfelty and Fromm 1996: Introduction, p. XVIII). Reading texts through this lens reassesses their reception. Adopted in numerous fields of research, including film studies, ecological approaches have opened up many fruitful avenues for study (Bouvet 2013). The turn of the twenty-first century saw the publication of two works analysing the ideological functions of environmental issues within Hollywood cinema: *Green Screen: Environmentalism and Hollywood Cinema* (Ingram 2000) followed by *Hollywood Utopia: Ecology in Contemporary Cinema* (Brereton 2005). This entry takes as its point of departure the influential collective work *Ecocinema Theory and Practice* (Rust et al. 2013). This work envisages cinema as a form of mediation that reflects ideological and cultural functions and suggests that any film can be analysed from an ecological perspective. As such, this position paves the way for a global ecological approach to cinema while calling for a distinction between ecocriticism and ecocinema.

The term "ecocinema" groups together a corpus of films that evidence ecological discourses or an environmental awareness. Outside of aesthetic considerations, it applies to categories such as ecological documentaries,[1] ecofiction, or climate fiction (cli-fi). The category of ecological documentaries includes both TV reports and documentaries. The term "ecofiction" is applied to films that explicitly address ecological issues, particularly cli-fi,[2] and which fall within cinema's broad generic categories,[3] such as science fiction. Ecofiction, a well-established and thoroughly researched subgenre that first emerged in literary theory, offers an ambiguous discourse, somewhere between denunciation and morbid fascination (Neyrat 2015), which can bring to light the multiple

DOI: 10.4324/9781003455523-20

and contradictory reasons why some mainstream American cinematographic productions stage the ecological emergency in an overtly spectacular way. Publications in the last decade have analysed a larger range of films – for example, Nordic cinema (Kääpä 2014), Chinese cinema (Lu and Mi 2009) – or embraced new multidisciplinary approaches but without necessarily considering new points of departure. Thematic readings have thus remained focused around the landscape, representations of wild untouched nature, or documentaries about activists.

From the 1920s onwards, cinema has been considered both as producing capital and as one of the rare forms of artistic expression that is able to reach the masses. Therefore, it is at once manipulated and considered as a tool of manipulation. This duality partially explains the need for American researchers to reveal the environmental discourses and counter-discourses present in the films produced by the Hollywood industry. For the same reasons, there is substantial temptation to overestimate the ability for films to influence public opinion. Moreover, a neo-Bazinian concept of cinematographic ontology (Bazin 1985) over-emphasises the notion of cinema being a disintermediated representation of wild nature associated with cultural factors.[4] This position identifies the *wilderness* and its corollaries as a systematic way to approach ecocriticism in film studies, thus limiting the corpus of films selected and themes undertaken.

The approaches developed by Stephen Rust can be exploited to a much greater extent. Many films that do not offer an ecological discourse or are not focused on representing nature can be analysed from an ecocritical perspective. For example, at first glance *Inherent Vice* (dir. Paul Thomas Anderson, 2014) may seem far removed from environmental issues: an outlandish Californian private detective investigates a disappearance. The film makes multiple references to film noir, including the neo-noir classic *Chinatown* (dir. Roman Polanski, 1974). According to David Ingram, the environmental dimension of *Chinatown* is diluted by its 1930s setting (the California water wars that inspired the film took place at the beginning of the twentieth century) and even more so by the revelations of incest that considerably detract from other issues in the storyline (Ingram 2000: 152–3). *Chinatown* and *Inherent Vice* both include scenes shot near the concrete canalisations that carry, or rather privatise, water, an issue where there is suspected mafia involvement in each film. The intertextuality in *Inherent Vice* thus allows for an environmental reading of *Chinatown* and brings to light the dark role of water in California. To make this role clear to the viewer, it must be shown in the film's imagery.

It is therefore essential to develop ecocritical analyses in a way that is comparable to ecopoetics, as outlined with reference to French literature in the introduction to "Littérature & écologie: vers une écopoétique" (Blanc et al. 2008: 15–28). Ecopoetics is an approach that is above all sensitive and attentive to the formal workings of film and to the capacity for films to raise awareness of ecological values through means specific to cinema. Analysing films, including those that fall outside of ecocinema, from an ecopoetic perspective, would reveal the diverse ways that environmental issues are conceived in order to direct viewers' attention toward these modes of representation.

Aside from some rare exceptions, ecocriticism within film studies remains close to the confines of academia. Nevertheless, studying films and their reception and reading them in line with the current state of an ever-changing world is precisely one of the purposes of film criticism, much more so than research. Without engaging in debates over the distinction between critical and academic discourses (Bergala 1996) and the substantial porosity between them, journalistic criticism can be described as "a daily clinic held by the bedside of the sick" (Sainte-Beuve 1862: 365); or, to cite Blanchot, "even if criticism is barely reflective, it comments, it interprets: it is turned towards the world. Its role is to draw the works out of themselves" (Blanchot 1992: 140). In this article, republished in one of the first editions of the film periodical *Trafic*, Blanchot invites us to consider film criticism in its capacity to draw the interiority of the work to the exterior, to an ephemeral present that is sick. Why then would bringing an ecocritical perspective to cinema not be one of film criticism's current concerns?

And yet, film criticism has been slow to incorporate ecological approaches, which makes the *Cahiers du Cinéma*'s recent initiative of dedicating their April 2019 edition to creating an herbarium of plants and flowers in film all the more special and unique. In his editorial entitled "Cahiers verts", Stéphane Delorme expresses the need to decentre our perspective, which is drawn to star actors, and refocus it towards the largely ignored vegetation:

> As for the critics (and that includes us), they most often talk of "trees" or "flowers". This is quite convenient because they do not really know what they are seeing. There is no power in attributing these labels. It reveals a profound disinterest in nature and an impoverished experience.
>
> *(S. Delorme 2019: 5)*

The issue's clear goal is to "make us see the world and films differently" in order to "fight against the impoverishing of our senses by an economic and political power that is only interested in reducing humans to consumers" (Delorme 2019: 5). This first attempt creates an herbarium in which plants, as a presence that contributes to the inner workings of the film, are named, shown, and reinvested with their role through brief analyses of films. These analyses often revisit heritage films or well-loved classics through the lens of a given plant to examine what it can convey metaphorically or narratively, especially in its visual power, its capacity to take over the image, redefine the space, and draw in human beings.

By consciously distancing itself from the types of films to which ecocritical discourse has thus far been confined, and by working sensitively in relation to imagery, this issue of the *Cahiers* breaks new ground for an approach that can be as prolific as it is promising, an approach capable of cultivating and enriching the reception of films.

Notes

1 Among those regularly cited are *Darwin's Nightmare* (dir. Hubert Sauper, 2004) and *Le Monde selon Monsanto* (*The World According to Monsanto*) (dir. Marie-Monique Robin, 2008).
2 For example, *Soylent Green* (dir. Richard Fleischer, 1973), *Snowpiercer* (dir. Bong Joon-ho, 2013), and *Mad Max: Fury Road* (dir. George Miller, 2015).
3 For example, drama with *Promised Land* (dir. Gus Van Sant, 2012) or the crime film *Bad Lieutenant: Port of Call New Orleans* (dir. Werner Herzog, 2010).
4 For example, Thoreau's influence on American visual culture combines a unique apprehension of the landscape that contributes to cultural density with the influence of the wilderness (see Mottet 1998).

Three Key References

Delorme, Stéphane (2019). "Cahiers Verts". *Cahiers du Cinéma* 754: 5.
Neyrat, Frédéric (2015). "Le cinéma éco-apocalyptique. Anthropocène, cosmophagie, anthropophagie". *Communications* 96: 67–79.
Rust, Stephen, Salma Monani, and Sean Cubitt (eds) (2013). *Ecocinema Theory and Practice*. New York: Routledge.

20

ECOFEMINISM

Frédérick Duhautpas

In the simplest terms, ecofeminism is the intersection of concerns within ecological and feminist thought. The term covers such a wide range of theoretical and activist approaches that it would be reductive to bring them together in one unified category. Consequently, this entry will not limit the concept of ecofeminism to a fixed definition, nor prescribe good practice for using the term. My humbler aim is to reflect upon the avenues that open up the intersections between ecofeminist thought and ecosophic perspectives. An ecofeminist approach implies a non-essentialist perspective which places two forms of domination at the centre of its analysis: the domination of human beings over nature and that of men over women. Its starting point is the assertion that important historical, social, and theoretical connections exist between these two forms of domination, particularly regarding the discursive mechanisms and materials that underpin and legitimise them. Incorporating the question of gender into ecological ethics means including the social and mental dimensions when examining the relationship between humans and nature. Ecofeminism thus offers a radical critique that throws into question Eurocentric, capitalist, and patriarchal structures.

The reason for including gender as an analytical category within an ecological reflection is not always evident. It is therefore appropriate to clarify where ecological and feminist thinking overlap, and which aspects could be applied to artistic practice. Within the framework of questions raised by Félix Guattari's mental ecology, which acknowledges the consideration of other ways of being (or more precisely "becoming", "occurring") beyond alienating existential constraints, this connection already plays out on a conceptual level, namely in interrogating the conceptual frameworks that justify and legitimise relationships of

DOI: 10.4324/9781003455523-21

domination. One approach within this intersectional perspective consists of highlighting and deconstructing the frameworks that condition normative ways of thinking, perceiving, and interacting with the (natural and/or social) milieu. According to the epistemological criticism proposed by feminist thinking, these relationships are founded on a dualist concept of differentiation and hierarchisation between the sexes. A conceptual framework is a discursive collection of beliefs and attitudes that condition and configure frameworks of perception, how we see each other, and how we see the world. Many conceptual frameworks are oppressive in nature, not only because they are reifying, but because they legitimise and standardise toxic modes of interaction that perpetuate relationships of domination and subordination.

Amongst these conceptual frameworks, gender (considered here exclusively in the singular) designates a system of bicategorisation founded on the oppositive relations of differentiation and hierarchisation between the sexes. Central to feminist thought is a deconstruction of this system's representative norms. The category of gender is based on dualistic thinking that essentialises individuals by drawing on the distribution of archetypal qualities that define identities, that is a machine that produces binary modes of subjectivation. This distribution is based on a reasoning that links the masculine/feminine opposition to other attributes that are also founded on a binary distinction (intellect/emotion, soul/body, culture/nature, strong/weak, objective/subjective, active/passive, interior/exterior, listening/speaking). As such, it perpetuates representative systems founded on these distinctions and associates them with relationships of hierarchisation, as one of the two terms generally becomes valorised to the detriment of the other. These dualisms establish a hierarchical relationship in which the inferior position is often reduced to that of an object and otherness, defined in opposition to the valorised position (here, the masculine position).

An ecofeminist approach extends this interrogation to our relationship with nature and to our milieu. Androcentric and anthropocentric norms, as well as the mechanisms of domination that they produce, rely on the distributional and hierarchising logics that emerge from the related conceptual frameworks. Ecological action also plays out in the symbolic register of the discourses, representations, and socialisation norms that influence individual behaviours. Numerous predatory behaviours towards nature, animals, or people have been culturally valorised as positive acts of conquest and a power of life ("*puissance de vie*"). Androcentric norms, especially that of the dominating male figure, have long been held up as models. Anthropocentric normativism emerges from the same thought patterns in terms of how it leads to a speciesist precedence of human beings over the rest of the living world. It is not by chance that the symbolism of patriarchal culture often associates the abstract image of nature with the idea of femininity and its accompanying essentialist archetypes and motifs (Mother Earth, fertility, reproduction, cycles, nurturer, care). In contrast, the masculine prerogative is associated with ideas of culture, civilisation, technology, the conquest of nature, consumption, and profit. The processes of overcoding and normalising integrated global capitalism combines with the mechanisms of patriarchal domination to establish a world founded on the exploitation, assignment, and submission of both women and nature. The intersectional declinations of ecofeminism also extends these questions to their interrelations with other forms of oppression (racism, homophobia, transphobia, ableism, classism).

The deconstruction of these dichotomies, notably those that rely upon the oppositions of listening/speaking, body/spirit, subject/object, self/other, is one of the numerous junctures between questions relating to ecology and feminist criticism. Deconstructing oppositive frameworks that associate women with nature and all the other qualities attributed to them is to revalorise both the idea of femininity, the inferior position within the social hierarchy that establishes gender, and our relationship to the world and to nature, often perceived simply as a passive deposit of resources at our disposal. Simply undertaking this type of questioning in light of these issues represents a distancing of oneself from those representations in favour of an ecological conception. The issues that ecology and feminist criticism address intersect here in terms of how they can be used to analyse the ways that discursive systems, whether communicational, commercial, or entertaining (publicity messages, soundscapes, visual landscapes), interfere with subjectivation processes, invasively influencing the construction of relationships to ourselves, the other, and the world. Ecofeminism involves paying attention to and becoming aware of how our milieu conveys orders for conforming to different oppressive social norms, thereby defining the standard way of being. Beyond political action on the environment and women's rights, ecofeminist criticism makes us aware of how our environment, or more specifically our milieu, favours certain modes of identification and dependence. As such, ecofeminism works to deconstruct predatory and oppressive ways of conceptualising alterity as a simple source of gratification. Instead, it aims to develop an ethics of listening that is able to consider its singularity, its balance, what it is becoming, its agency, and the reciprocal relationships that it leads to.

An ecofeminist ethics in art would thus propose creating sensorial spaces that encourage the development of a consciousness that allows us to eschew oppressive frameworks. This process involves developing ways of feeling, thinking, and acting that are more conscious of our relationship with others and with our milieu. No programme, recipe, or procedures are specified; it is simply the idea of developing other ways of thinking about our relationships beyond the mechanisms of identification based on dichotomous distinctions (man/nature, man/animal, man/woman, culture/nature) and their related processes of reification. This approach is linked to a criticism of the normative patterns that can also enter into artistic creation. For example, within sound creation practices, approaching sound ecology from a desire to deconstruct the dichotomous presumptions and logics on which relationships of domination and reification rest may intersect with methods of feminist musicology. This approach revalorises listening in our relationships with the other and the world. Traditionally, artistic practices have often underestimated listening in favour of speaking, understood here as the auto-centred expression of the demiurge artist delivering his message to the world. Cultural perceptions that are often attached to listening are imbued with the dichotomies of passive/active, interior/exterior, and subjective/objective, invariably valorising one term to the detriment of the other. Perceived as a passive activity of receiving (culturally coded as femininity and dependence) through its opposition to speaking and the act of enunciation (culturally coded as masculinity and power), listening suggests a form of obedience. The practice of sound ecology involves abolishing the duality between listening and doing, precisely by centring its activity on listening. Reconnecting with active listening also means becoming aware of how the voices that surround us in our immediate environment can affect us and prove to be oppressive. It is about using the senses to become aware of how the logics of domination in the milieu are linked to one another and are made possible by means of a sustained exposure to the orders that divide up our living space. It also represents a search for a balance between the actor and the acted upon by deconstructing the oppositions between interiority and exteriority, the subject and the object, which are often linked (in terms of the categories men/women, human/nature), in favour of a conscious interrelation of equilibrium. Paying attention to the self is to promote listening, empathy, and understanding as qualities and attitudes that are no longer considered essentialised feminine attributes, but as the essential elements for developing a balanced relationship to the world. Simply put, these approaches aim to develop a concept of subjectivity that can extract itself from toxic frameworks, thereby promoting a higher standard of well-being for ourselves, others, and our living space.

Three Key References

Delphy, Christine ([1991] 2001). "Penser le genre. Problèmes et résistances", in *L'ennemi principal 2: Penser le genre*. Pari: Syllepse, 243–60.
Plumwood, Val (1993). *Feminism and the Mastery of Nature*. London: Routledge.
Warren, Karen J. (2009). "The Power and the Promise of Ecological Feminism". *Multitudes* 36(1): 170–6.

21

ECOFEMINIST TERRITORIES

Tiziana Villani

Many of the most recent forms of artistic expression are connected to activism and different strands of contemporary feminism. These practices and creative approaches to research address corporeity, spaces, territories, and thus ecological questions understood in the most comprehensive meaning of the term and not simply "environmental". From the Arab Spring to incursions from milieus once considered outside of the paradigm of Western centrality, forms of expression are emerging that offer a range of actions and produce "cracks" or artistic "fractures" that, temporarily at least, overturn the reigning acritical aestheticism of the artistic act. Art invents its own "minor language", subverted in relation to the dominant mainstream.

Accordingly, ecofeminism has broadened the spectrum of its actions and theoretical reflections by drawing on art, often as a favoured medium, in all its diversity. Today's feminist struggles have plunged traditional feminist movements and many of their historic past demands into crisis. The body, gender, and identity can no longer be analysed from a binary perspective. Similarly, ecology goes beyond an exclusively environmental preoccupation, which is only one part of it. More holistic approaches can interrogate the logic of patriarchal domination within the work itself as well as the intersection of social categories like gender, race, and class, and questions of colonialism in the fight for land rights. For Silvia Federici (2004), this represents the inherently conflictual dimension of the commons as processes that are sparked by struggles.

The genealogy of what can be defined as a materialist ecology is strongly linked to forms and practices of critical thought. The concept of "natureculture", developed by Donna Haraway (2003), explores environments within the social, political, and technological domains that examine processes of resistance in relation to the ever more devastating power of dispossession within modern existences.

DOI: 10.4324/9781003455523-22

The ways in which art contributes to defining more accurately the shape of these struggles and forms of resistance are written by the body, a body that reinvents itself through flash mobs, performance, and often ironic renderings of the supposed reference models for the dominant narrative. Yet, in the activism of numerous countries in Latin America, Africa, and the Indian subcontinent, these are always the bodies and subjectivities that question the withdrawal of a colonialism that still produces geographies of war. Ecology cannot ignore these challenges that are often posed by communities of women. As the writings of Gloria Evangelina Anzaldúa show, intermediary spaces or those where cracks appear that the capitalist model still wishes to penetrate recount other territories that offer opportunities for escape. Nepantla, a Nahuatl word meaning intermediary space, identifies temporal, spatial, psychic, and intellectual points of liminality and potential change. Within the concept of Nepantla, individual and collective notions of self and visions of the world shatter. Supposedly fixed categories, whether based on gender, race, sexuality, economic state, health, religion, or a combination of these elements and often others, begin to erode. Borders become more permeable and start to crumble.

By inhabiting states of Nepantla, we are transformed into what Anzaldúa calls "nepantleras": mediators, in-betweeners "who facilitate passages between worlds". Anzaldúa coined the word "nepantlera" to describe people on a boundary, on the threshold: those who inhabit more than one world and who develop what she describes as "a perspective from the cracks" (Anzaldúa 2009: 322). Nepantleras use cracks between worlds to invent holistic, relational theories and tactics that allow them to rethink and transform the different worlds in which they exist.

Today's feminisms explore these fractured territories, the thresholds beyond which it is possible to abandon codes, norms, and identity constructs. These transitions can be considered open bodies in the sense indicated by Haraway. In other words, they are performative grafts that unite technology with the animal, that invent alliances and generate kinship in a less narrow way than in traditional institutions.

Alterity is displaced in these strictly ecological transformations, in these entirely political movements. When alterity manifests itself – and I would say that with technology its manifestation is disturbing because it accelerates, intensifies, and modifies the very concepts of emotion, affection, and strength – contemporary societies are compelled into a rather worrying reflection as it represents what must be standardised.

Yet, how is it possible to standardise a field made up of corporeal, mental, imaginary, and fantastical protheses? By exhausting all the potentialities that open up with a virtualisation of existence, a virtualisation that spreads, extends, and develops in the direction of the norm, Anzaldúa used communication strategies that would become fundamental. Some of the authors already mentioned have a significant interest in hybrid women and cyborgs, interpreting them in specific ways. Why would the cyborg have more significance than other forms of feminine subjectivation? Why is it easier to connect the woman and the animal? Queer theory offers some useful insight, like Daniela Daniele's preface to an Italian essay collection *Meduse cyborg* (1997), which references Judith Butler. The different examples cited move beyond the traditional understanding of a cyborg as part-automaton, part-human, part-animal:

> Judith Butler's *Bodies That Matter* and Donna Haraway's female cyborgs go through the female body to transcend biological limitations. As in the case of the cross-dresser or the shaman, the woman-medusa's body emerges stronger by assuming the full powers of masculinity and femininity, like Diamanda Galás's amplified and multi-vocal usage of the voice (*über-voice*) or Laurie Anderson adopting a masculine voice tone by using an electronic instrument, the vocoder. Just as performance reinvents the limits of art, these women's bodies appear volatile, permeable, and queer, offering the chance to reformulate their own destiny beyond common understanding and thus producing unusual paradoxes.
>
> *(Daniele 1997)*

In sum, becoming-woman evokes an ecosophy that has no defined goal, single objective, or moment of revelation. That makes it an anti-philosophy: race, gender, affiliations, identities are not abstract categories in which and from which dominant narratives are articulated. Being aware of them is not enough; opposing these articulations is not enough: we must find ways to liberate the imagination that experiment with other situations and conditions, whilst recognising that micropolitics can also take different and often contradictory directions.[1]

The technologies and the transitions underway call for resources in line with the accelerated change we are experiencing. If this change is left to formulate a unique horizon of command at will, it can only inflict new "thorns", to reference Elias Canetti, that mark and will mark new processes of human selection and, more generally, living beings. Spatiality or the territories are thus to be considered as true fields and bodies of experimentation, but they are also always open and mutant bodies, political, decolonised, and exploratory, which can readily be employed, rather than sad passions that tend to humiliate them (Canetti 1962).

Note

1 I do not have the space to examine micropolitics in depth here, but it is certain that minorities have a clear relationship with the forms of micropolitics that Guattari discusses at length (Guattari and Rolnik 2007).

Three Key References

Anzaldúa, Gloria (1987). *Borderlands/La Frontera: The New Mestiza*. Portland: Aunt Lute Books.

Guattari, Felix, and Suely Rolnik (2007). *Micropolitiques*. Paris: Les Empêcheurs de penser en rond.

Villani, Tiziana (2019). *Corps mutants. Technologies de la sélection de l'humain et du vivant*. Paris: Eterotopia France.

22

ECOSOMATICS

Marie Bardet, Joanne Clavel, and Isabelle Ginot

The term "ecosomatic" refers to a field of study and a set of practices that reject any separation between the body and its "others", and which develop a sense of self as a milieu for the other living beings whose presence makes our own life possible.

Somatic Practices

Somatic practices consider the body as soma, an indivisible physical, emotional, and mental corporeity that is inseparable from its milieus. These practices emerged in the Western world towards the end of the nineteenth century and first flourished in the 1920s. Rather than an entire discipline in and of itself, they constitute a hybridisation of multiple knowledge sources, a bricolage of practices and theories from different areas: Western scientific and medical knowledge, sports movements and productivist concerns, new pedagogies. Somatic practices developed against the historical backdrop of Europe in the process of environmental, industrial, and urban transformations. The circulation of knowledge from non-Western worlds, such as yoga, meditation, or martial arts, relies on colonial dynamics whose power relations shape the hybridisation of knowledge that cultivates alternatives to representations and usages of the body in society (Guilbert 2011; Launay 1996; Hanse 2010; Locatelli 2019; Boumédiene 2019). The second rise in somatic practices primarily took place in California in the middle of the twentieth century and was characterised by cultural exchanges with non-Western worlds. The wealth of practices, including the Feldenkrais and Alexander methods, Eutony, Body Mind Centering, Rolfing, and Life Art Process, develop a set of techniques

DOI: 10.4324/9781003455523-23

that hone perception as a source for the quality and diversity of movement. Somatic practices are a relational form of pedagogy, often based on constructing situations to be explored, which aligns them with improvised dance practices and other contemporaneous pedagogies. Through exploring constructed situations, these practices produce a wealth of experiential knowledge about how humans can experience the world and how this experience is actively shaped by shifts in socio-environmental habits and conditioning. Current usages are marked by an individualism. Patenting practices as products places them in the context of a well-being market economy. Designed to remedy urban ills and the hyper-productive lives of the middle and upper classes, this recontextualisation bypasses their rooting in radical social and political criticism and the emancipatory issues that have been integral to them since their emergence. These practices, after all, were part of the utopian social and political movements in the 1920s and 1960s. And, despite an epistemology that remains marked by anthropocentrism, they share common concepts with ecology and hold the very real potential to become a laboratory for a cosmopolitical intervention project that shares knowledge with other than humans.

Identifying Issues to Better Prepare for Them

With the term "ecosomatic", we aim to extract somatic thought from the liberal logics on which it depends and to highlight its convergences with other critical and political dynamics. The intention is not to create a fixed category or define a rationale, but to outline and understand a move towards collectively thinking around four types of increasingly urgent issues:

- Creating a body of knowledge composed of, first, theoretical and experiential knowledge developed through somatic practices and, second, scientific and political knowledge of ecology that obliges us to simultaneously think about temporal scales (histories of human beings in nature and those of intertwined civilisations) and spatial scales (from that of the human being, whose activities have an impact well beyond their living environments, to that of Earth system science).
- Not separating environmental questions from social and political questions. In other words, it is a case of capturing the dynamic and evolving balance between the planet and living beings in questions of subjectivation, corporeity, sensations, movement, and the imaginary, and of considering these processes not as private and individual, but as singular and collective, as proposed by, among others, Guattarien-inspired ecosophical thinking.
- Developing a politics of knowledge that considers practices and types of knowledge that are experiential, micropolitical, subjective, indigenous, or

minority as legitimate as mainstream knowledge, which will no longer take precedence.

- Developing a political engagement that considers work linked to somatics (teaching, research, practice, creation) as a political and ecological project invested with responsibilities towards the processes of emancipation, resistance, and creation that surround us. We must ask whether the usages of somatic practices go hand in hand with the affirmation that our theoretical actions are practical actions integrated into disputed fields and ways of being in – and making – the world.

The Body as Nature: Materialities, Histories, Representations, Reopening Trajectories

Ecological thought, like movement or dance studies, has moved away from any reference to nature or the body as being stable, substantial, ahistoric, unequivocal, and universal categories. Somatics opposes a corporeal reductionism that views the body as a "living machine" and divides the subject into specific anatomical parts. Instead, it offers a relational, processual, and dynamic way of thinking about living beings. Modernity or, to follow Harraway, the Capitalocene submits the body to a politics of remote power. It violently controls the reproductive strengths of women's bodies in the same way as a colonised labour force. The notion of a "disciplinary" and disciplined body remains dominant even though numerous philosophical and political currents have deconstructed it many times over. For example: the critical thinking on movement that tends to undo the Cartesian dualism still haunting Western thought; the trans-feminist and queer currents dedicated to deconstructing a substantialist and essentialist biologism; and the situated theoretical practices emerging from postcolonial concerns. Somatics extracts itself from this reification of the body through different approaches:

- Relational thinking where the body is understood as a dynamic of relations and modulations within a milieu.
- A multiplicity of actors (fascia, muscles, bones, cells, organs, bodily fluids, respiratory organs, energies) that work together in a multipolar articulation with different thresholds of consciousness and attention to the subject. This view introduces a diversity and deconstructs hierarchies at the very heart of the subject by distributing roles to other cellular or tissue entities, even imaginary ones.
- Perceptive and multimodal work that seeks to deconstruct the hierarchies between the senses, particularly the domination of the visual and the concealment of the kinaesthetic.

Art Worlds

Somatics constructs a wealth of resources that are at once practical, technical, and conceptual. It permeates numerous artistic practices engaged in a critique of the hierarchies and elitism of the learned arts and supports the study of art as an experience of feeling in touch with the world. Artistic works and processes employ the same modulations of feeling, listening, and sensing as somatics in order to "melt" into the world, with the aim of understanding it better and transmitting this presential and relational quality to the environment, thus experiencing the porosity between oneself and different environments (natural, urban, or social). Similarly, projects described as "participative", contributory, dedicated, or situated that transform non-dancers, everyday people, and non-specialists into artistic actors also draw on sensory techniques that are accessible to all. Anna Halprin was a central pioneering figure of these choreographic approaches that abandon traditional virtuosity to engage the senses in new ways of inhabiting environments (natural, urban, social), and raise social and political questions (including relations of racialisation, gender and sexuality, or health). As such, somatic practices overlap with numerous artistic approaches including situated choreographies (Perrin 2019), sensory walks (see "Walking Art" in this collection), or immersive projects that explore the concept of art as an aesthetic experience through knowledge of perception, rather than as a spectacle encapsulated in representation.

Three Key References

Bardet, Marie, Joanne Clavel, and Isabelle Ginot (eds) (2019). *Écosomatiques. Penser l'écologie depuis le geste*. Montpellier: Editions Deuxième Époque.

Clavel, Joanne, and Isabelle Ginot (2015). "Pour une écologie des somatiques?". *Revista Brasileira de Estudos da Presença* 5(1): 85–100.

Ginot, Isabelle (eds) (2014). *Penser les somatiques avec Feldenkrais*. Lavérune: Éditions L'Entretemps.

23

THE GARDEN

Gilles Clément

If I were to offer a synthesising definition of the garden in just a few words for a traditional reference work, I would say: "The garden, land of hope for the mind". This phrase is not all encompassing, but it gives an essential summary of what the garden represents in my opinion. To plant a seed in the ground is to invest in a future where one hopes that the plant chosen to germinate and grow will produce flowers to look at and bear fruits, leaves, or roots to eat. It is a gift for the future. There is no space for nostalgia, only hopes and surprises. With a garden, very often, one discovers the unexpected; one is rarely disappointed.

The garden's original purpose was to provide sustenance. Yet, over time, its function changed to become a means of decoration. One does not preclude the other. A garden is an architectural type of decoration where form and construction in the space take precedence. Different styles, linked to different cultures, reveal ways of seeing beauty at its best, what must be protected at all costs. Hence the word "enclosure" translates as "garden" in certain languages.

The Garden Today

The perilous state of diversity leads us to consider every garden as a space that can accommodate and nurture at-risk species, assisted only by the soil and the climate. Beyond form or decoration, the garden is a living environment where the behaviour of the species planted there never aligns with a fixed vision for the space. Ephemeral plants are vagabonds who take advantage of biological opportunism, often dependent on climatic whims, to stray across the terrain.

To protect species, it is enough to respect their behavioural variations. Today's gardener no longer carries out plans in the manner of a geometrician or an artist. Evidently, making artistic decisions forms part of this ephemeral

DOI: 10.4324/9781003455523-24

work; but the gardener is becoming the benevolent assistant of the vegetation, animals, fungi, and micro-organisms that live alongside plants. The term "weeds" – literally "bad grass" in French – is disappearing from our vocabulary. No grass is bad, only potentially badly situated. It may be removed from the place where it is a nuisance to the plants that need protecting, but it must live elsewhere in the garden. Eradication techniques are also disappearing. There is no reason to develop them further in a global context that seeks to balance a complex ecosystem of living beings, rather than promote a devastating monoculture.

Under these conditions, the garden becomes a place of dialogue with beings whose behaviours we know little about, but which we must respect to assure our own survival. We depend on diversity because we exploit it. This is the prerogative of living beings, a biological privilege of our territories, not to be subjected to functionalist, formal, or competitive demands. This attitude shifts away from an approach that wastes unnecessary energy cleaning up the garden to one that is no longer chained to a hygienist or profit-orientated vision.

The current crisis demonstrates the formidable resourcefulness of living beings in an unbalanced world. A micro-organism can impose the laws of biological opportunism right up until the point of curbing its energies in favour of rebalancing ecosystems. The garden based on the prerogative of living beings creates an unstable and dynamic collective where equilibrium is always readjusting itself.

Three Key References

Clément, Gilles (2015). *The Planetary Garden and Other Writings*. Philadelphia: University of Pennsylvania Press.

——— (2012). *Jardins, paysage et génie naturel*, Leçons inaugurales du Collège de France 222. Paris: Fayard. URL: http://lecons-cdf.revues.org/510; DOI: 10.4000/lecons-cdf.496

Clément, Gilles, and Jamaica Kincaid, et al. (2023). *Garden Futures: Designing with Nature*. Weil am Rhein: Vitra Design Museum.

24

GEOGRAPHY AND THE AESTHETIC PRODUCTION OF ECOLOGICAL ISSUES

Joanne Clavel, Clara Breteau, and Nathalie Blanc

It is a tough challenge to shape or embody ecological crises. Whether it be climate change, biodiversity loss, pollution, or land degradation, it has proven particularly tricky to shape these into concrete and sentient forms. In other words, it is difficult to produce aesthetics for these crises. There are multiple reasons for this: (i) aesthetics finds itself delegitimised at the negotiating table as scientists view it as subjective heresy; (ii) aesthetics is not yet recognised as an academic discipline up to the task of engaging with these issues and thus has less contact with ecology and its implications; (iii) artists or philosophers working on aesthetics show little awareness or concern over industrialisation and the devastating effects of current lifestyles in terms of their mediated and technologised aspects; and (iv) the languages of maths and computing (statistics, modelling, big data) are the only means of politically representing nature in the world of governance (local and international), in the media, and for the general public.

Nevertheless, Baumgarten's founding definition of aesthetics positions the discipline as a fundamentally ecological science before its time. Aesthetics is an awareness of the sensory elements that form an integral part of the environment, a knowledge that encapsulates the interweaving relations which shape each object's milieu. Aesthetics is configured as another way to give form to what surrounds us by representing as accurately as possible the constitutive relations that are central to the apparatus of sensory knowledge. Before aesthetics was sidelined and swallowed up by the notions of art and beauty, before becoming defined by objects rather than as a mode of knowledge and relation to the world, the perspectives it originally offered were epistemological and environmental.

DOI: 10.4324/9781003455523-25

Over the course of the twentieth century, numerous conceptual tools were developed that cultivated the idea that the emotional is political. Authors such as Debord, Foucault, Deleuze, Guattari, Nancy, Agamben, and Rancière equipped generations of academics and artists to denounce "the society of the spectacle" (Debord) and its unequal and alienating "distribution of the sensible" (Rancière), the structural biopolitics that marginalise and oppress, as well as the necessity to reinvent plurality and new futures over and over. Since these ideas and discourses have not yet succeeded in gaining momentum on a mass scale, they may still be too inconvenient, utopian, or unsettling. Perhaps they also remain too anthropocentric and fail to take a real interest in Mother Nature, this curious "witch" who manages to speak without words.

Since the beginning of the twenty-first century, French landscapers and geographers have been developing aesthetics as a way of thinking about nature and milieus that integrates emotional experiences of environments and places (urban, suburban, rural, or wild). In the field of geography, two avenues have been identified: first, the work of Berque (1987) reconsiders the notion of milieu from the perspective of phenomenology; second, the work of Blanc (2008; 2016) anchors the environmental aesthetics developed in the USA and the UK (Berléant, Brady, Carlson) in a theory of environmental and everyday interdependences both material and ideal (Godelier 1984). The works of Anne Volvey (2007) and Pauline Guinard (2016) sketch out a geography of art and emotions. At the same time, several environmental humanities researchers have identified ecological crises as being principally crises of the emotions and of experience, attributing a central place to aesthetics in their works (Bardet et al. 2019; Clavel 2017; Morizot 2016). These attempts interrogate, for example, the way that living milieus and autonomous lifestyles participate in the production of poetic qualities in places and habitations, giving rise to new vernacular cultures (Breteau 2018).

The relationship between geography and ecology, cousin disciplines envisaged in the nineteenth century as the studies of milieus, has undoubtedly become tumultuous and problematic. Indeed, from the 1950s onwards, geography was built on a fear of environmental determinism that led to the ecological perspective being excluded from the field for several decades (Lévy and Lussault 2003). Moreover, geography's concern with gathering knowledge about the world, its regions as well as their material, environmental, or cultural specificities is linked to humanity's project of mastering the environment. As Besse (2003) has shown, multiple apparatuses of geographical knowledge are based on a cognitive programme that aims to represent the Earth in terms of a scopic and overarching rationality. Yet ecological crises urgently require an ostensive definition of the world built on entirely different types of knowledge, experience, and ideology. Without being exhaustive, four approaches can be identified which can help geography, once engaged in a new "relational paradigm" (Chan et al. 2016), to participate in the aesthetic production of ecological issues: research-creation, questions linked to experience, relationships between art and urbanism, and articulations of scale.

First, collaborations between artists and geographers that produce new tools (e.g., emotional maps) contribute to the development of relational representations between individuals, collectives, and living environments (Mekdjian and Olmedo 2016). New practices are invented in the processes of creation, production, and diffusion, in the different ways of doing art and doing science. Carbon consumption, waste production, and the use of extractivist technologies and creative materials are reconsidered in favour of manual, collective skills conceptualised for specific places. Artists integrate themselves into urban spaces and collaborate with residents to champion socially engaged artistic or scientific processes. Approaches to research associated with environmental aesthetics promoting research-creation practices are being sketched out. Within these practices, a new schema is developing based on the recognition of place as a singular situation and on the co-construction of a study and the objects of research. It is a question of combining means of geographic representation, aesthetics as a practice associated with a multitude of experiences, and environmental ethics as conditions for the functioning of living collectives formed by humans and non-humans. Subjectivities and corporealities once again inhabit places and their representations.

The role of geography in the aesthetic production of ecological issues is also evidenced through the importance, in relation to pragmatist philosophies, of the resident's and the citizen's experiences as political actors who acquire a capacity for initiative and transformation (Dewey 1934a, 1934b). Encounters with the alterity of other beings in nature and integration into a situated milieu to which we create attachments reinvent the landscape as a historic battleground that connects and reconnects the human and non-human inhabitants of a neighbourhood, village, or valley. The lived experience of places thus becomes a double lever that encourages mass mobilisation. It shapes and sets the narratives and emotions that drive political action. These forms of landscape are thus made visible and tangible, which effectively integrates them into a new distribution of the sensible.

Geography is also at the forefront of studying the increasingly close relationship between urbanism and the arts. While a rising number of territorial collectivities place artistic intervention at the heart of urban transformation (renovation, appropriation of land, infrastructure planning), the role of these aesthetic producers is very different depending on the context. In the northern suburbs of Paris, the cultural cooperative Cuesta and the theatre company Gongle set up a "game space" that drew on the principle of sports match commentary. By creating a space where residents could express their hopes and fears about the urban environment, its atmosphere, practices, and usages, an artistic project provided a pool of resources for the architects and urban planners working in the area.

Finally, due to its interest in articulations of scale, geography can become a space of reinvention for a movement that brings together studies on the destructive effects of globalisation,[1] a reconsideration of imaginaries and projections about our shared world, and practices that develop local knowledge. We must learn to unlearn. It is a re-education that begins with listening to what exists, the temporalities of living beings, and thinking about the smallest movements. Limiting human action on the planet requires us to invent other forms of being in the world and of presence. This is a wakeup call, an invitation to invent the most outlandish imaginaries of ecological (not environmental) aesthetics that practice and think in step with what inhabits it and what it inhabits.

Note

1 This is a million miles from Syvlie Brunel's climate-sceptic work.

Three Key References

Blanc, Nathalie (2008). *Vers une esthétique environnementale*. Versailles: Quae.
Breteau, Clara (2022). *Les vies autonomes, une enquête poétique*. Arles: Actes Sud.
Clavel, Joanne (2017). "Expériences de Natures, investir l'écosomatique", in *Le souci de la nature, Apprendre Inventer, Gouverner*, Cynthia Fleury and Anne-Caroline Prévot (eds) Paris: CNRS, 257–69.

25

GRAPHIC DESIGN

Yann Aucompte

The evolution of the branch within French graphic design that is steeped in issues of social ecology will be of interest to those seeking to understand questions of the environment in a work's production. Numerous graphic designers are developing activities that have a truly ecosophic effect, according to Félix Guattari's definition, on their environments. Even if these approaches do not directly address nature, there is, within them, "ecology", understood as a sense of the complex interactions between different configurations of reality, a desire for a complete reconstruction of political, aesthetic, and mesological ecologies.

Through working in public spaces, graphic designers are giving new purpose to an overused medium: the poster. The most famous posters function as "paintings for the street" and can be found in museums and galleries alongside paintings. However, this does not naturally suit the medium, which has a particular place in the history of graphic design and whose visual dimension was only developed relatively recently. Nowadays, the poster's visual appearance is overemphasised, its informational purpose forgotten. The word "poster" has largely replaced the historical term "placard" despite the latter very much representing the lineage with media forms from antiquity, such as dipinto and stone inscriptions. Placards are a central part of a history of writing in the public sphere in France. During the French Revolution, for example, posters took on a stronger civic role as they served to communicate with the people. Still in a textual form, posters then accompanied the nineteenth-century revolutions and the Commune. For commercial reasons, posters became more visual and colourful in appearance from 1850 onwards. The works of contemporary politically engaged graphic designers that involve introducing an individual's words into the public sphere are therefore the heirs of the placard tradition, which pre-dates illustrated posters. For graphic

DOI: 10.4324/9781003455523-26

designers, this represents a return to reading posters and their images as texts in the public sphere. In today's world, which is saturated with images, it is a shift away from "reading but unconsciously retaining" (Perrottet 2012: 127).

This socially conscious trend developed in a particular way from the work of a group of graphic design auteurs. These types of practices were immortalised by the collective Grapus, who were influenced by the organisation of Ateliers populaires in 1968. Grapus, which split into two factions in the 1990s, outlined distinct ambitions for left-leaning graphic design. In fields from education to culture, its members have attempted to bring their approaches to institutional objects, for example designing the Louvre Museum's visual identity (Pierre Bernard). Moreover, graphic designers can embrace symbolic ecology in complex ways. By positioning themselves against advertising, they have tried to invent a mode of polysemic communication. In some compositions of signs no monological interpretation can emerge; rather they require interpretation and a capacity to adapt. Grapus's other heirs focus on developing graphic design projects in disadvantaged environments, for example, working in prisons, for literacy organisations, and in working-class areas. Some make a living facilitating cultural activities, providing writing services for the public (*écrivain public*), or engaging with inhabitants as a "graphic designer in residence".[1] Within these activities, the authoritative role of the graphic designer disappears to allow local actors to produce their own visual environments with a strange familiarity.

Far from positioning the "designer-as-hero", socially conscious graphic designers seemingly proclaim: "we may have to imagine a time when we can ask, 'what difference does it make who designed it?'" (Rock 2013: 54). These graphic designers shine the spotlight on productions by communities whose voices they represent. The production is "exhibited" in libraries, cultural centres, and in the streets. Often, they stage a performance or create an event from the production itself. Graphic designers take advantage of the design processes' capacity to immerse them in a specific place and thereafter work on its environment and the event, rather than on an artistic gesture or act. Working in a participatory manner with the inhabitants transforms a convivial instance of working together into a collective moment in which the production once again becomes poietic and practising artistic techniques sheds light upon all the relationships of production within the "social pedagogical approach" (Paris-Clavel [2018] 2022).

Graphic designers are generally linked to the associated milieus, drawing from them language, relationships of production, and ideas. Whilst artists often mask these links so that it is ultimately the inalienable autonomy of humanisation that is performed, graphic designers are caught in a heterogenesis. To paraphrase Michael Bierut, the designer designs in a more adapted way as he or she learns from their subject. There are thus two fundamental aspects of ecosophic practices or those described as such: they are collective and permeable. These practices form institutions that do not fall back on their apparatus but remain open to transformations. Accordingly, in our age's regime of governmentality and its liberal economic discourse, they become dependent, servile, and invisible. The fact of this disappearance of socially conscious graphic design makes it a truly ecosophic practice, conforming to Guattari's definition, an endless becoming in minority.

From these activities, graphic designers and those who collaborate with them (residents and citizens) create a collective work that is undefinable – in the same way that aesthetic theory has difficulty defining it (Passeron 2000). There is no author in the sense of a rational subject with the desire to facilitate things around them, making them "master and possessor" according to Descartes's expression (Afeissa 2007: 95) of the world. Instead, a collective subject and productive meetings constitute a political or authorial body that is undefined in terms of legal identity. Writer, sociologist, artist, bookbinder, administrator, volunteer, and activist come together and join forces on projects. They work in equal parts on productions alongside workshop participants. The category of author, as a rational and paternal individual subject, disappears. More than an author, it is a collective in which "there is a constitution of subjectivity at a scale that is transindividual from the outset" (Guattari 2013: 41).

This definition of aesthetics and human nature recalls Dewey's affirmation of the anthropological aspect of artistic emotion, understood not as being specific to viewing works in a museum, which reveals a "class exhibitionism" (Dewey 1934c: 6), but as being as much about the artist as the worker who perceives the materials as he or she works with them. For Dewey, aesthetic emotion is nothing to do with the privilege of art and the artist, but "the intelligent mechanic engaged in his job" (1934c: 5), who finds aesthetic emotions in technical handiwork. The conditions of production (workshop or factory) and of aesthetic appreciation (museum or supermarket) very often prevail over all other factors in aesthetic judgement.

By escaping the art world and its clinical spaces, graphic designers re-enter life. During the intervention, they release actor-citizens from systems of industrial production and re-establish local practices. Though their work only relates to the images and signs of our environment, they propose a new model of individuation whose relationships of production can call for a greater understanding of permaculture, principles of local debate, and attention to the environment. In doing so, they reduce mesological harm and implement virtuous approaches of economy that respect the environment and nature – without even these ideas being the subject of their work, without even their names being proclaimed in the name of art.

Note

1 In the mould of "architect in residence" (*"permanence architecturale"*), a practice popularised by Pierre Bouchain in the field of architecture.

Three Key References

Guattari, Félix (2013). *Qu'est-ce que l'écosophie?*. Paris: Lignes/Imec.
Passeron, René (2000). *Pour une philosophie de la création*. Paris: Klincksieck.
Pater, Ruben (2021). *Caps Lock*. Amsterdam: Valiz.

26

I FOR ICONŒMIC

Giusy Checola

The word "scene" designates the fixity of a bordered visual space (e.g., the scene of a crime) and the flux of urban life (e.g., the arts scene). As a visible enactment of social phenomena, it expands the space of everyday life in terms of a "sense of scene", "an arrangement of elements (people, actions, things, places) in which a certain social or moral condition is expressed" (Casemajor and Straw 2017: 16). Similarly, iconic public art projects and sculpture as infrastructure, particularly those that provide access to essential resources, create what I term "performativity of place", understood as the capacity for the "created" place to improve, increase, guide, and even control the perception and experience of reality. Such projects and infrastructures have the potential to affect two forms of memory that constitute "the basis of our identity" and our knowledge about the world: first, explicit memory, which is conscious and analytical, guides our physical and virtual movements; second, implicit memory, which is experiential and secretly imprinted on our bodies, guides us in an automatic and all-encompassing way. Relational memory, which depends on collective semantic memory, establishing the contexts in which we live and influencing the individual stories we compose from our lives (Lejeune and Delage 2017: 9), may also be affected.

Implicit memory manifests "like an image on a photographic plate whose surface must be carefully examined to make out the contours" (Lejeune and Delage 2017: 9; see also 8, 99, 111, 113). The *"image-calcul"*, which Monique Sicard refers to, and computational imaging more broadly, simultaneously call for an ontological and photographic framework and the "impossible visualization of the network", an integration of and a disconnection from its inherent nature of being "out-of-frame", that is both its context and its borders. Since

DOI: 10.4324/9781003455523-27

the *image-calcul*'s matrix is at once "mother and mathematics", "genesis and classification", it thus reveals "a common part between each of us, engendering sensorial shifts in our shared references and our visual cultures" (Sicard 1998: 9–10). In consequence, an iconic and infrastructural art of place also acts as a potential strategy, screen, and device for addressing memory, just as the city and the computer act as what Ranjodh Singh Dhaliwal has called "addressing machines".[1] In other words, they act as an automatic addressing system with collective meanings and functions.

However, let us put forward two propositions. First, we can consider images as forms of life that "intervene in the imaging process, without any intention from a transcendent author" by means of "a perception between images, occupying the same iconic environment 'beyond the human', trans-spatial and trans-temporal" where "the images 'eco-speciate themselves'" (Durafour 2017: 15–16). Second, we can acknowledge that since the Byzantine Empire the Western Eurasian economy has been intimately linked to the visibility of the icon and the creation of "a repetitive and fertile visual world" by an iconographer, "where the mirror is the invisible quiddity of being, *not represented* because not representable" whilst "what is shown, rather, puts in place the visible formula of something that will ensure the stability of an empire" (Mondzain 2005: 155). Thereafter, we can affirm that the iconic character of an art of place has the potential to ensure the visibility and livability of the formula of stabilisation or disruption of dominant forms of power. At the same time, it allows for unprecedented combinations of visual imprints and sensorial experiences of place through contravening previously accepted rules. This provokes "a paradigm leap" (Arecchi 2011: 66) that affects fluctuations in the imaginary meanings of landscapes within the objective environment, reproducing the "inherent ambivalence" of the modern landscape (Berque 1999: 154–5) and its existential condition of cultural image and physical place (in the sense of Cosgrove and Daniels 1998).

As the method and concept of "place-making" were emerging in the United States from observations of social behaviour made by urban planner William H. Whyte and the writer and activist Jane Jacobs to address security and poverty on the outskirts of cities and to rehumanise the design of shared spaces, the Italian geographer Eugenio Turri attempted to define these imprints with the term "iconema". Iconema refers to "the iconographic transposition of the notion and function of the phoneme" that acts as an "elementary unit of the perception of the landscape", particularly in relation to the Italian landscape, which, in combination with others, determines the image of a place that a community "instinctively refers to each time a landscape or specific place comes to mind" (Turri 1979). How can the image be reduced to an elementary unit like a phoneme, especially if it results from the iconographic, physiological, and sensorial transposition of perceptions of a landscape? And how, in consequence, can it not be susceptible to further "segmentation", particularly if some of its "distinctive traits" are common to several cultures and if they are often the result of different forms of remediation (in the sense of Bolter and Grusin 2000) and cross-border manipulation across the infosphere, the realm of migration, archiving, and the management of data memory?

Developing an "iconœmic" methodology for studying art as a "place-maker" – understood as the creation or regeneration of foundational relation-ships between the main imaginary meanings of places (in the sense of social creation, see Castoriadis 1997), their geoaesthetic experience as mediated by art, and the construction of different symbolic, primary, and social infrastruc-tures – would allow for the redefinition of the site-specific concept of "iconema" as the more complex and interdisciplinary "iconœma", thus connecting the creation of iconic places to "intersensorial economies" (Casemajor and Straw 2017: 4). In this formulation, the "œ" vowel is emblematic of the intrinsically multi-faceted character of the iconic image of place as a device that influences and orientates experience and cognitive processes. As the "œ" constitutes a digraph (a formula of linguistic coexistence), typographically a ligature (a new and specific form whose spelling is considered as having an etymologising effect), and a diphthong grapheme, it goes beyond and overrides the borders of Romance languages.

By giving form, infrastructure, and image to collective imaginaries and sen-sorial experiences of place, the symbolic-phenomenal functions of art as place-maker are thus considered as essentially "performing" and "modelling" in nature.

Note

1 Dhaliwal used this phrase during the online roundtable "Cities and Computers: Our Urban-Machinic Imaginaries" hosted by the UC Davis Humanities Institute in 2020. Dhaliwal's research on addressability has subsequently been published (see Dhaliwal 2022).

Three Key References

Arecchi, Tito (2011). "Dinamica della cognizione. Complessità e creatività", in *Paesaggi della complessità. La trama delle cose e gli intrecci tra natura e cultura*, Roberto Barbanti, Luciano Boi, and Mario Neve (eds) Milan: Mimesis Edizioni, 55–88.

Cosgrove, Denis, and Stephen Daniels (1998). *The Iconography of Landscape: Essays on the Symbolic Representation, Design and Use of Past Environment*. Cambridge: Cambridge University Press.

Mondzain, Marie-José (2005). *Image, Icon, Economy: The Byzantine Origins of the Contemporary Imaginary*, trans. Rico Franses. Stanford: Stanford University Press.

27

LANDSCAPES, TERRITORY, AND URBANISM

Alberto Magnaghi

Territory[1] is a collective work of art and science built over the centuries in a "choral" form (Becattini 2015). It is humanity's greatest form of expression in history, transforming *lands* into *landscapes, natural spaces* into *places*. Territory is born out of human efforts to fertilise nature. Together, they grow in a coevolutionary process to produce new living ecosystems. Territory acquires a more complex identity through the cultural *médiance* (Berque 2000) of the civilisation that produced it. When a civilisation declines and dies, territory also declines and dies (deterritorialisation) to be reborn anew with the next territorialisation. With each rebirth (reterritorialisation), territory becomes denser, deeper. The new civilisation destroys some traces and geographies of the previous territorialisation and incorporates other traces and artefacts, culturally reinterpreting and integrating them into the contemporaneous landscape. This cycle continued until the dawn of a machine-based civilisation, our Anthropocene, which initiated a separation of nature and culture. Once the coevolutionary process between human settlements and the environment was abandoned, territory as a living system was destroyed and replaced with an artificial and hyper-technocratic urbanisation of the planet (megacities), which, independent of nature and history, is characterised by eco-socio-catastrophic and entropic tendencies of human settlement (new forms of poverty, low urban living standards, pollution, devastating effects of climate crises on the biosphere, pandemics).

In spite of this destructive path of structural deterritorialisation where the art and the science of coevolution no longer underpin the production of space, strong indications of the ongoing construction of land and landscapes persist in our everyday living environments. Territory, although wounded, still exists.

DOI: 10.4324/9781003455523-28

This existence value (resistance, resilience) of territory allows me to propose a utopian and optimistic vision of a possible renaissance as part of a transition to an eco-territorial civilisation that can restore the value of its living heritage.[2] Above all, this vision depends upon this value being understood, reinterpreted, and represented simultaneously as a *scientific* and *artistic* work.

We can consider it as a technical-scientific work because each period (e.g., Etruscan, Roman, high and low Middle Ages, the Renaissance) develops technical knowledge to enact the coevolutionary process between human settlement and the environment in terms that favour the human milieu with its regulations, styles, and reproduction statutes. These areas of knowledge include hydraulics, geology, ecology, climate, agro-forestry, cartography, urban planning, infrastructure, and construction. Despite being organised into specific disciplines, these forms of knowledge directly and indirectly interact and cooperate with each other towards the successful (technical and social) "choral" production of space in its territorial uniqueness, its ability to differentiate places, its landscape identity, and its capacity to reproduce itself.

We can consider it as an artistic work because the structures underpinning the technical-scientific construction of territory in its historical development imprint on the Earth's surface morpho-typological traces that radically change its original identity. These include living environments, urban and rural landscapes, and infrastructure which interpret "the art of building" – culture, regulations, and knowledge – through the architectural and urban works of the current territorialisation. These traces have accumulated layer upon layer over the course of several civilisations to create the complex landscapes on which we walk and live today, the landscapes that we cross, observe, breathe in, and listen to. This extraordinary work of art, perceptible through all the senses simultaneously in our "life-worlds", has been produced collectively by the successive generations who have created unique landscapes in every corner of the globe. These "*tableaux vivants*", each on a unique palimpsest, recount the synthesis of the historic territorialisation process in artistic form, from nomadic cultures to the subsequent transformation into settled civilisations up until modernity.

Today, we are faced with destructive processes that are causing a radical deterioration in the quality of our urban and rural living environments. How then do we restore a coevolutionary (artistic and scientific) construction of territories, cities, and landscapes?

The task is very difficult; however, being historically a matter of collective, "choral" art and science, it can draw its origin and strength from a simple revival of inhabitants. In other words, inhabitants must be returned to the centre of decision-making processes and reinstated as a collective entity in new forms of self-government of cities and territories.

The progressive dismissal of inhabitants as a collective figure that constructs and cultivates their own living environments, and their transformation into customers and consumers of goods on the global market, are the basis of the decline in the quality of territory, milieu, cities, and, indeed, of the landscape's beauty. Reversing this perspective requires an immediate acceptance of the goal to put the inhabitants' territory as the subject of knowledge and design first, to be open to Pierre George's (1993) proposed shift from the man as a producer to the man as an inhabitant.

This process of reconstructing the inhabitants' territory calls for:

1 A collective knowledge of the peculiar identities of places through a reactivation of the relationship between expert knowledge (aesthetic-perception, morpho-typological, structural approaches) and contextual knowledge of memory (inhabitants' cultures), using tools that encourage participation and self-representation (community mapping) for collective care, with an increase of "place-consciousness".[3]
2 The capacity to reconstruct collective relations of proximity, public spaces, and forms of community, by drawing on tools for grassroots planning where the inhabitants are the main players in decision-making processes concerning their life-worlds (self-government) and their landscapes "as they are perceived by people" (Council of Europe Landscape Convention, Article 1a).
3 Reconnecting forms of residence and consumption with local economic systems in order to ethically bring about social and environmental well-being through an enhancement of territorial heritage as a common good managed by inhabitants/producers in sustainable and renewable ways.
4 The revival of the classic dialogic principle described by Leon Battista Alberti (the relationship between the architect and the other – experts and members of society – across forms of pedagogy, consultation, and approval), which Françoise Choay (2004) sums up: "there is no construction without dialogue with those for whom one builds: individuals and communities made up of members of the family and of the *res publica*". This principle becomes all the more important if we extend it from the building to the city, to territory as a complex place of contemporary habitat, which has the value of common good, to which active citizenship should be applied in the different forms of participatory planning (Marson 2019).

This path towards the centrality of inhabitants in the production of space, through a new "art of building" on the regional (territorial) scale, defines the scientific and artistic tools for the project of urban bioregions (Magnaghi 2014) that are able to restore coevolutionary relationships between human settlements and the environment. This vision opens up ecological aspirations for beauty and quality of living in new landscapes as common goods of our life-worlds.

Notes

1 Following the Italian "Territorialist School", I adopt here a definition of "territory" as "a highly complex living system [...] produced by the long-term", coevolutionary, and synergic "interaction between human settlements and the environment, cyclically transformed by successive civilizations" (Magnaghi 2005: 62). Therefore, all the following occurrences of the term and its derivatives here should be read with this meaning in mind.
2 The Italian Territorialist Society brings together researchers from different disciplines, who are socially active in research/action projects, to work on renewable local development projects with the aim of restoring territorial heritage. See: www. societadeiterritorialisti.it.
3 "Place consciousness can be defined as the awareness, acquired through a process of cultural transformation of inhabitants/producers, of the patrimonial significance of territorial commons. The shift from the individual to the collective defines [...] the reconstruction of elements of community in open, relational, and shared forms" (Magnaghi 2017).

Three Key References

Berque, Augustin (2014). *Poétique de la terre, essai de mésologie*. Paris: Belin.
Choay, Françoise (2011). *La terre qui meurt*. Paris: Fayard.
Magnaghi, Alberto (2020). *Il principio territoriale*. Torino: Bollati Boringhieri.

28

LEARNING AND EXPERIENCE

Anastasia Chernigina and Antoine Freychet

In the face of ecological imbalances and a lack of appreciation for sensory experience in our relationships with our environmental, social, and technical milieus, it would be judicious to reconsider artistic practices (in transition) and (ecological) learning together. In both, the action of a subject in a given situation, the lived experience of that situation, and processes of constructing meaning are linked. There is neither learning nor aesthetics without an experiential process; conversely, there is no experience without cognitive and affective processes. An ecological artistic education, or artistic eco-training, thus seeks to take place as close as possible to reality, including and drawing on commonplace situations in life and the everyday relationship between the subject and things, other beings, and themselves.

Local and Contextual Knowledge, and Situation-based Learning

To know the world and what constitutes our embodied existence of it, we must extend our ways of understanding our environment, adapting to it, and prioritising relationships that are in resonance with the world (Crawford 2014; Rosa 2019). This knowledge necessitates experience and the trial and error, detours, toing and froing between moments of autonomy and cooperation that are inherent in it. The more specific, intense, unexpected, or touching the experience is, the more chance we have to gain from it: the skills to pose and resolve problems, to react appropriately to unexpected events, to open oneself up fully to a given situation (Boumard 1996). Then, we must acknowledge the fact that learning also, and perhaps even above all, concerns our "pre-reflective consciousness", preconceptual as well as "concrete memory". Learning is

DOI: 10.4324/9781003455523-29

connected with the orientations of our psychological mechanisms and our perception, and to sensory-motor functioning that forms the general backdrop of life, without which one would not scratch the exact spot that a mosquito, whose acoustic presence provoked flashes of exasperation, bit (Récopé et al. 2013). Situation-based learning increases these advantages.

Certain artistic practices lead participants to discover, activate, and/or, in effect, refine their sensory perception. For example, the goal of workshops that combine somatic and listening exercises is to become aware of the sounds in our environment and our corporeal sensations. To notice what is there through experimenting with listening to and observing sensations, soundscapes, and the physical environment, participants become capable of detecting new phenomena (e.g., broom seeds being ejected from their pod in the crisp air, small bran grains dispersing in time and space amongst the bushes) and new activities (a beetle crossing a dead leaf, a market gardener on a tractor ploughing a potato field), and guiding their listening, their movements, and their imagination in a different way, according to what they encounter.

Holistic Learning

This type of learning has a holistic dimension. Emerging from a concrete and situated activity, it draws on the inseparability and interdependence of the cognitive, affective, conative (motivating the passage to action), and corporeal dimensions at play within lived experience. Situation-based learning requires time and regular practice, but it allows the subject to recognise their mistakes and the consequences of their actions. Once open to exploring our sensorial environments, the learner encounters non-intentional objects and learning. The learner develops skills and aptitudes that were not specifically envisaged, which decentres the relationship to learning by removing it from voluntarist, authoritarian, or market logic.

This can be illustrated with the example of sound. As a relational, affective, and cognitive medium, sound constitutes a link between the environmental milieu, as it is manifested in the architecture and the dynamics of place, and an individual, as it affects one's thoughts and feelings, produces memory, links memories, and mobilises cognitive capacities to the extent of inserting itself into the very structure of the subject's living system. Listening is a unique way of knowing that entails attention to detail and to the unexpected. This acuity goes hand in hand with different types of learning relating to our social existence (since we are connected to other living beings, inorganic elements, places, and the particularities of each one), things (empirical knowledge, examination of meaning and origins), and the self (one's place in the world, behaviours, responsibilities). This is what makes learning *from* listening and *through* listening profoundly ecological, encompassing the mental, social, and environmental dimensions of ecology.

Situating Knowledge

Connecting pedagogical, ecological, and aesthetic questions allows us to formulate a criticism of neoliberalism in which aesthetic experience, one's relationship to the world, and learning are employed to accumulate capital and not for their emancipatory potential. We propose to develop situated knowledge within an educational framework of critical consciousness. Rather than simply transferring or transmitting knowledge and values, education must be a cognitive act (Freire 1970). This type of education begins from problems, necessities, and criteria specific to the cognitive subject (to their body, their history, their "sensitivities towards") and tends to preserve the link with the situated dimension of knowledge and judgement. In what conditions and for what end is such knowledge produced? What exactly does it lead to, and what does it imply (Haraway 1988)?

The critical impact is accentuated by the fact that becoming conscious of material and technical, social, and political situations can lead to new possibilities for collective action. By understanding the world, the self, and the commons as realities in transformation, the learner increases their own empowerment; they are empowered in line with what surrounds them. Their awareness of and abilities to participate in their environment in a holistic way, based on what is necessary, relevant to them, and what they aspire to, are strengthened. This offers the potential for real transformation for the betterment of the self and the world.

A Collective Dynamic

A truly ecological education, which would include a social dimension, must be accessible to all and strive for emancipation. The individual would be left to formulate their own opinions on ecological issues based on experience and act in accordance with them. The aim is to develop an autonomy of learning. Instead of being independent of the environment, this autonomy would be gained through the subject refining their relationship with what surrounds them. Thereafter, we can share our positions, concerns, and worries regarding a given situation, whether local or global, discuss them, and, if necessary, seek collective ways of addressing them.

Developing a specific artistic experience with others creates a new form of communication and socialisation. During collective sound and somatic workshops, a listening community forms: time and place, sound phenomena, and atmospheres are shared. Orientating our listening experience is done through interacting with others, and this form of socialisation is connected to the particularities of each of us, non-human beings, and inorganic elements. It is thus a form of interactive learning that does not necessarily require verbal communication, rather it is transmitted through listening to the environment of which each participant feels they are an intrinsic part.

Circulating and Sharing Knowledge

An ecology of learning must reconsider the circulation of knowledge and skills. Rather than hierarchising access to knowledge, know-how, or tools, it is a question of circulating or sharing them. In doing so, learning benefits from this sharing and cooperation, developing a social and collective intelligence.

In workshops and places of residence or artistic production, circulating competencies is indispensable: for example, knowing how to involve people with different interests and expectations in a shared activity within an educational context; sound artists sharing low-tech solutions; and video makers discussing the question of artistic licences. A truly alternative network of artistic ecological practices can emerge from this approach. Moreover, it poses the question of balance and coherence between the various concerns of the project, particularly when there is a desire to appear ecological, and the technologies, energies, and financial means it requires. This is made possible by networking and sharing the specialities and energies that have an impact on that balance.

Conclusion

From a pedagogical perspective, combining artistic practices and learning facilitates the development of ways to encourage others to take responsibility and to reinforce and introduce affect in learning; to make it important, concrete, and situated, attentive to what surrounds us; and to combine the acquisition of practical tools and strategies that can be applied to contemporary life. A critical ecology approach leads to multiplying the perspectives to challenge dominant norms, entering into contact with worlds that are unknown to us, paying attention to places and non-humans, or even questioning our relationship to technology. In short, it develops truly ecological forms of consciousness, knowledge, and action. Considered in line with artistic approaches, this convergence opens up the possibility to connect with the flow of affects that circulate in everyday life and in learning situations, and to transform sensory experience, all the while making it available for others and the world. In other words, the concerns of arts in transition and ecological learning converge.

Three Key References

Crawford, Matthew B. (2014). *The World Beyond Your Head. On Becoming an Individual in an Age of Distraction*. New York: Farrar, Straus and Giroux.

Freire, Paulo (1970). *Pedagogy of the Oppressed*. New York: Continuum.

Haraway, Donna (1988). "Situated Knowledges: The Science Question in Feminism and the Privilege of Partial Perspective". *Feminist Studies* 14(3): 575–99.

29

LITERATURE AND THE COMMONS

Rémi Astruc and Thierry Tremblay

In *Commun*: *Essai sur la révolution au XXIe siècle*, philosopher Pierre Dardot and sociologist Christian Laval predict:

> The world will not be protected by establishing a sort of reserve for "common natural goods" (land, water, air, forests, etc.) [to which we must add literature, arts, aisthesis …] "miraculously" protected from the indefinite expansion of capitalism. All activities and regions interact with it. So, it is not so much about protecting the "goods" essential to human survival than profoundly transforming the economy and society by overturning the system of norms that are now threatening humanity and nature in a very direct way.
>
> *(Dardot and Laval 2014: 13)*

Their book-manifesto broke new ground in the French intellectual landscape by opening it up to reflections on the sharing of resources in a finite world, as introduced by Elinor Ostrom (1990), Nobel Prize winner in Economics. Dardot and Laval are outraged that the future is being confiscated by the tragedy of non-commons that is emerging from the contemporary cataclysm of economic war and ecological crisis. In response, they call for all disciplines to wake up.

Despite the book beginning with quotations from Walt Whitman and Charles Baudelaire, little place is given to literature, art, music, or any form of aesthetic experience (ethico-aesthetic paradigms according to Guattari), which are excluded from the ecological reflection and struggle for the commons from the outset in favour of the economic and political paradigms. Literature, the literary arts, one could conclude, have a strictly ornamental function. Their sole power appears to be epigraphic, to adorn as a chapter heading, which

DOI: 10.4324/9781003455523-30

lends a "spiritual" touch in a unique literary language that resides completely outside of the struggling world. Conceptualising literature in this way is truly an error. To save ourselves and the world from the forthcoming disaster, we can no longer isolate the sphere of exchanging goods in an attempt to reason with it. We will never escape the stupidity (and culpability) of humanity by believing we can take refuge in an exclusively aesthetic place that is miraculously spared from the malicious impulses swarming everywhere else.

When Pina Bausch cries "Dance! Dance! Otherwise we are lost" in Wim Wenders's documentary, she does not mean that it will be necessary to save dance from the disaster. Yet, it is only through dance, the arts, artistic practice, and other "creative" practices that disaster can be averted. In the founding principles of artistic action, generosity, which is lacking in the organisational and utilitarian world that makes calculations on existence, can be found. Besides, saving dance alone would not mean anything. Dragging beauty alone from the catastrophe would mean a sterile, useless, and to be frank definitively ugly beauty. If utilitarian art reveals itself to be useless, it then follows that art which is not at all utilitarian would be just as bad. Art can only be considered as life-saving when, as it is peripheral to the utilitarian sphere and calculating thought, it finds its true ministry in the general ecosystem of relationships: of humans to their environment and of the environment to its humans.

It is not by chance that Bataille's anthropology thus makes excessive action – expenditure, rather than economy – the act of balancing the system itself, the homeostatic protector of a liveable and sustainable life. This expenditure is manifested in artistic action, but also in war, as Tiqqun (2009), for example, reminds us. This strange war has inspired art since at least the time of the Lascaux cave paintings and other enchanting myths right through to the discovery of the poetic form. Bataillean expenditure is a useless expenditure that is, at the same time, directly useful for the accounting system within relations of expenditure. In other words, it is necessary to know how to spend to enrich oneself, to waste to save, and to lose one's head in order to think. This is the great anthropological lesson of artistic action, which re-establishes it as one of the forms that is absolutely necessary to balancing the exchange system of beings: communication (pooling exchanges, sharing identities).

In his *Abécédaire*, Gilles Deleuze (2012) offers a negative yet fair definition of the writer. A writer does not write about his grandmother's cancer, but extracts himself from his personal situation to write beyond his "little personal affair". If not, writers are what Deleuze calls, in a very conceptual manner, "stupid fools" (*connards*). The world has been devasted by "the abomination of literary mediocrity", and the world of literature is full of stupid fools who write about the death of their grandmother and henceforth will write about the death of the planet in the same individual and miserable way. Taking as a pretext the personal drama that one feels when confronted with the collective catastrophe we are destined for, opportunists get upset, protest, and take up the pen. Profiting from the disaster, these bandwagon intellectuals, or what Vieillescazes calls "*intellectuels d'ambiance*" (2019), have adopted the death-of-the-planet, death-of-the-species, or even death-of-mankind position because it presents an opportunity for them to become writers, to feel like or to believe they are becoming politically engaged writers. The reason why the system also needs this fringe of parasitism to both maintain and kill itself off at the same time is what allows us to understand the true nature of what a system is. As for the question of identifying who reads these parasites and why they are read, we will leave it to future generations to provide an explanation.

Effectively, there is no ecological turn in literature, no transition towards a new form of art, that is more respectful, more ethical. There is only an art of stupid fools that remains an art of stupid fools even after injecting some ethics and becoming a little greener. In contrast, there is a true art that has always had an immense connection to the community of living and non-living beings at its heart, which is only positioned as true art because it already ties that richness of relationships to the world to its centre. There is no "care" turn that literature could take; there is only a type of literature that cares about trends and cares about pleasing in an attempt to maintain itself and exist by parasitising once again a world that would almost certainly be better off without it. Authentic literature, in contrast, is an act of tearing, and because it tears apart, it can be the only thing that matters – and that saves. Literature does not stitch cuts, nor repair lacerations. But through tearing wounds apart, literature is empowered to transform the pain into a surplus of sensations. This is why literature does not have to concern itself with the commons, nor other things for that matter. It either is concerned or it is not. If it is, literature is already part of the commons, emanating from them and produced by the world through which it speaks to us and associates us with it.

The commons are, in this sense, the interior arrangement of the exterior, in the form of a community of bodies and thoughts, of material community and immaterial community whose link is fundamentally aesthetic. Community therefore means: the words that you say to yourself and those that you hear are the same words that, through sharing, participate in the commons of the community; and those same words also originate from this arrangement and are addressed externally, to the community, thus appropriating sensations and phenomena. The commons and ecology are thus essentially linked. Both participate in the same field and have the same root: οἶκος. The process by which ecology is integrated into the commons is οἰκείωσις, the fact of appropriating itself, and it naturally opposes alienation. The transition from a familial *eco-nomy* to a more community-based economy passes through the most ancient Christian theology, that of Paul the Apostle. When the economy extends itself to the whole ecumene, it merges with ecology, a relationship with the totality of what is habitable.

In our era of late-stage modernity, where the three ecologies are reversed, subjective appropriation of the totality as private ecology denies the medium of appropriation and therefore denies the commons by the community instruments themselves whose language is at once the most prominent and the most invisible. Subjectivity is enclosed in the social fabric, being in a group, which is itself planted in the often-forgotten environment. Yet this subjectivity is freed from the fabric and the elements in which it floats.

In other words, community is found between an apparently unsolvable subjectivity and exteriority, the shared becoming of elements of subjectivation. The community must be understood in its essence as a negative community, as what escapes all characterisation. It is not a return to a golden past or a celebration of a glorious present. It is the very space of signification and therefore the very space of the future, that of "real Territories of existence" (Guattari 2000: 35).

To sum up and conclude, the idea of community does not only designate an arrangement, a configuration, an (economic, sociological, psychological, etc.) agreement, an ethos that one can consider, evaluate, and integrate. Community also designates "of the inside" which is found "outside of": an aesthetics, an emotional way to orientate oneself, a sympathetic immersion in what is elementary, a form of attention or gaze, a communion between affects and the world that they interrogate, that they effectively attempt to inhabit and that they already effectively inhabit. This is how the great utopians are designing the community to come, as a community already invested in desire: a desired community, a desiring community, a community of desire.

Note from Rémi Astruc. It was with great sadness that I revised this entry alone. Thierry Tremblay passed away on 5 November 2022 in Prague. His thought-provoking work is a reminder of what his loss represents for us all.

Three Key References

Astruc, Rémi (ed) (2015). *La Communauté revisitée/Community Redux*. Versailles: RKI Press.

Dardot, Pierre, and Christian Laval (2014). *Commun. Essai sur la révolution au XXe siècle*. Paris: La Découverte.

Ostrom, Elinor (1990). *Governing the Commons: The Evolution of Institutions for Collective Action*. Cambridge: Cambridge University Press.

30

LITERATURE(S)

Aline Bergé

As a written and oral form of art linked to the diversity of languages and cultures across the world, literature has always borne witness to human societies' keen sensory and kinetic interest in the wonderment of living beings and their relationships to their living spaces and environments. This interest is manifested in: calls, movements, and gasps; handprints on cave walls and footprints in the snow; the oriole's song under the cover of a wood, transcribed by an unknown ancient Chinese poet. These sounds and inscriptions are all signs and manifestations of what is offered and taken away, at first spontaneously, in the exchanges between interior and exterior, conveyed through the breaths and rhythms of basic life, emotions, and the flow of bodies and materials. In search of traces and clues of the living world, literatures accommodate, translate, and circulate calls and mysteries, enigmas and marvels, as well as questions: "who sings when every voice is silent?" (Jaccottet 2014: 153).[1] Through a range of shape-shifting forms, literatures represent and compose the genesis, the plurality, and the interweaving of real and imaginary, natural and supernatural worlds. Whether animist, totemist, analogist, or naturalist (Serres 2009), literatures in different places and at different times recount the dawn and the dark, what has been and what will become of communities, how they live and cultivate the Earth, how they structure their relationships to others – gods, demons, other peoples, animals, vegetation – within the complexity of living beings. Literatures produce secular and sacred hymns and poems, cosmogonies and epics, myths and fables, stories and accounts, tragedies and comedies, historical investigations and riddles, songs, essays, and performances.

Literature evolved from ritual uses of language in the social, economic, cultural, religious, political, and ecological domains of human activity: rites of foundation, communion, and commemoration; rites of alliance, war, or

DOI: 10.4324/9781003455523-31

negotiation; rites of passage and initiation combined with seasonal rituals (van Gennep 1981). The simultaneously communal and individual experience of literature brings to life a poetic creation that is always situated, embodied, and dialogical, indissociable from an opening up to the world and an experience of alterity: an ecopoiesis of the "connection" between humans, land, and gods, self and other, here and elsewhere, the visible and the invisible, the sensory and the cerebral (Jaccottet 1968). Through the diverse spectrum of characters and alliances with the other in the poem, plot, subplot, or collection of stories, in dialogues or the polyphony of theatrical scenes, a "poetics of relation" is invented (Glissant 1990). This poetics is open to transformations, hybridisation, and the emergence of other possible worlds: an ethics of becoming other with others and a politics of transitions.

Using memetic, dialogical, and critical modes, literatures have long employed a framework of primary (Pinson 2020) and secondary ecologies inherited from past and recent intersecting histories of the Earth and human societies, the specialisation of writing, and the division of knowledge to serve the powers that be (Goody 1977). Literatures, as with the arts in general, are governed by an ever ambivalent and paradoxical relationship with the scientism of modern and western Europe and the new empires of world capitalism (Graeber 2015). Despite being creative in nature and responsible for collective memory and patrilineal, matrilineal, and inter-species heritage, literatures are bound by and subordinate to social hierarchies and market values, which have exiled them from the political sphere and confined them to entertainment spaces. Yet their interest in the commons means literatures have long warned of: violence, partitions, and instability in the world; the conflicts and forgotten moments in history; the theft and exploitation of land and human beings; inequalities and social and environmental injustices; and the deprivation and privatisation of access to vital resources as part of globalisation (Raharimanana 2008). As witnesses to the past century's disasters, to the multiplying attacks on living beings, and to human societies' exposure to increasingly extreme risks, writers continue to compile accounts of life and survival in toxic environments (see Alexievitch 1998, who would go on to win the Nobel Prize in Literature in 2015). Writers respond to loss with the gift of a "book of healing", like Assia Djebar's homage to Avicenna in her acceptance speech at the Académie française (2006). Outside of dystopias, from the turn of the 1970s onwards, there has been a rise in new indigenous and vernacular literatures, documentaries, and ecofictions, which reconnect humans to nature, to animals, and to vegetation, led by the lively and humorous pens of ecofeminists, from Monique Wittig (1969) to Le Guin and Ursula (1989) and Donna Haraway (2016). Concurrently, ecocriticism, introduced by William Rueckert (1978), invites readings that reveal the "stored energy" and the generative, and thus native, power that the "ecosystem" of a poem holds.

Both art and lifelong companion, living memoir and inventive matrix of the commons and individual transformations, literature offers a path to becoming sensitive to, raising awareness of, and initiating environmentally friendly ways of acting that respond to and attempt to resolve the ecological crisis. Going against the grain of death cultures and, like Joséphine Serre in *Data Mossoul* (2019), defying the new empires of digitalisation and artificial intelligence that drive humans to distance themselves from the real world and reduce arts, writing, sciences, and living bodies to the "knowledge of a stockkeeper" (Florence Dupont 1994), writers and artists are increasingly interweaving human existence and knowledge of living beings into their practices. Their collective calls to action and for mobilisation embed literature back into the political sphere: Édouard Glissant and Patrick Chamoiseau directly addressed Barack Obama in the name of humanity and "the uncompromising beauty of the world" (Glissant and Chamoiseau 2009),[2] whilst other writers joined with civil society to take the climate emergency to parliament in order to examine the capacity for humans to reduce their hubris and act within the planet's limits (Patriarca, 2015). All of these attempt to integrate sensory intelligence about the state of the world, humanity, and interdependences alongside the power of utopianism, imaginaries of the future, and desires for a common horizon (Chamoiseau and Le Bris 2018).

To paraphrase Hölderlin: why living literatures in times of ecological crisis? We must understand and know how to maintain the fragile balance of equalities in the work, in the act of reflection, and in the vital leisure of literatures, arts, and sciences. We must further connect the work of ecopoiesis to a concern for the Earth and caring for humans; to the collective invention of spaces of coexistence and increasing hospitality; and to citizen debates and defending common assets. This is the "cultural battle" that André Gorz (2019; 1967) called for. Buzzing with the chatter of thousands of languages, in a world that is today mostly plurilingual and ever more mixed, nurtured by memory and new experimental and scientific knowledge about biodiversity and the climate, literatures convey the stories and the imaginaries that humanity and the world are searching for. The tangible and intangible cultural heritage of humanity that literature and the arts offer has only recently been recognised (UNESCO 2001); the history of humanity and global commons is just beginning.

Notes

1 Many thanks to John Taylor, a contemporary translator of Jaccottet, for his insights and providing the translation of this quotation.
2 For Glissant and Chamoiseau, "in*trait*able" in the original French title (*L'intraitable beauté du monde*) has a double meaning. The term alludes to a revolt against or resistance to the "*Trait*e des Noirs" (the slave trade); in other words, a resistance to slavery.

Three Key References

Colombo, Fabien, Nestor Engone Elloué, and Bertrand Guest (eds) (2018). *Écologie et Humanités*. Special Issue of *Essais, Revue interdisciplinaire d'Humanités* 13. URL: http://journals.openedition.org/essais/411 (accessed 1 December 2019).

Rumpala, Yannick (2019). *Hors des décombres du monde: écologie, science-fiction et éthique du futur*. Ceyzérieu: Champ Vallon.

Schoentjes, Pierre (2020). *Littérature et écologie. Le mur des abeilles*. Paris: José Corti.

31

MEMORY AND CHOREOGRAPHIC WORKS

Isabelle Launay

How can art form connections between history, works, and ecology? Within the Western tradition of theatrical dance, the persistence of the work, in spite of the vagaries of oral tradition across time and memory, has long been envisaged in terms of a linear heritage, a thinking that "preserves" works independently of any circulation in the milieus of the present. At the turn of the twenty-first century, the passing of numerous pioneering choreographers (Martha Graham, Merce Cunningham, Pina Bausch, Rudolf Nureyev) raised the issue for their dance companies of how their repertoires would survive and provoked reflection on questions of their revival, re-enactment, and transmission. Susanne Franco and Marina Nordera open *Ricordanze* with the phrase "against the ephemeral" (2010: XVII) to deconstruct this topic in dance studies. Under which conditions can a tradition, a repertoire, or a choreographic work, whether so-called "classic" or "contemporary", be extracted from the network of conservative forces in which it might be held? Can a work be passed on without suppressing its creative dynamic? Can the same action both save a work and throw it into crisis to preserve as much of its current relevance against the consensus of what is already known? In other words, can a work be protected from commodification on the market of choreographic repertoires?

To illustrate this possibility, I will analyse the case of les carnets bagouet, named after the prolific French choreographer Dominique Bagouet (1951–92).[1] Over a 16-year period from 1976 to 1992, Bagouet created more than 45 pieces, around 10 of which are considered major pieces in contemporary choreography in France (Ginot 1999), with a permanent company that brought together almost 80 dancers. Dying prematurely from AIDS, Bagouet could not or did not willingly want to discuss what would happen to his work after his

DOI: 10.4324/9781003455523-32

death: he left the question entirely open. His close collaborators established the organisation les carnets bagouet in Montpellier in 1993 just after his death. Due to their relationship to Bagouet's work, les carnets bagouet strongly risked being caught up with problems linked to the dynamics of heritage, such as being conscious of "fidelity" and the susceptibility of transmission to repression. It was thus necessary for them to overcome difficulties relating to the feelings of possession and affective transfers formed during those 16 years of work, then to organise the conditions for a potential appropriation by any member where each would take responsibility for its direction.

More or less conscious of these risks in 1993, the group knew that choosing a name was already reinventing its heritage. les carnets bagouet thus stood firm in their desire to be with *and* without Bagouet, of proposing their own project in the name of the choreographer whose memory they evoke. The name was chosen, not because Bagouet left notebooks (*carnets*), but to refer to the collective: according to the performer Catherine Legrand, "these carnets, it was us, it was the people".[2] The notebook is the very object in which a link between the sensorial memory of the dancers and its relative historical distancing manifests.

Following the selection of a name, a horizontal and non-hierarchical collective structure was created, which sought to be light touch and, rather than expressing the necessity of reviving and maintaining a repertoire, affirmed the need to discuss it. The practice of contradictory democratic debate was placed at the centre of the project: working together over time on different points of view, desires, and modes of being present within a collective was therefore one of the significant challenges they faced. It was a case of not simply choosing that heritage but inventing how to create it and at what rhythm: being able to participate, withdraw, and return according to the necessities of the present. As the dancer Matthieu Doze wrote in 1997:

> We chose to conduct this work together. [...] It seemed evident and right to us to form this collective, each taking some responsibility in this common project. [...] finally, it's still the desire [...] to construct together that history whose foundations we know but the rest is to be invented and reinvented. [...] We all talk about the same thing, but we talk about it in different ways. This formidable richness alone justifies our coming together.
>
> *(Letter from Matthieu Doze to Christine Lemoigne,*
> *cited in Launay 2007: 146–7)*

The functions of the performer multiply and the borders of work are redefined or displaced: one project lead becomes choreographer, another assistant, another takes on a variety of roles such as pedagogue or performer, all in accordance with the needs and responsibilities that each individual accepts to take on. les carnets bagouet thus seems a unique case in the history of dance and so-called contemporary theatrical dance specifically. Never before has a permanent company of performers on their own initiative taken charge of a renowned choreographer's work after his death to establish the conditions of a transmission relevant for that particular repertoire following an alternative eco-micro-political model that avoids as far as possible the future promised by the aesthetic-liberalism model of repertory companies.

In the collective's view, a work by Bagouet can only be revived under two conditions:

1 It must be performed not by dancers (who can always be replaced), but by people who dance, or even by teenagers, amateurs, or taken on by other contemporary performers following specific work plans and desires, thus prolonging the approach of Bagouet himself.
2 Each revival must be able to work with different frameworks of transmission, depending on the work and the context.

Through collective self-analysis, les carnets bagouet have sought to:

• Resist calls for a fixed repertoire and examine the need of whether or not to keep the memory of a work alive;
• Free themselves from their past experience as "dancers" of what they called "the-dance-of-Dominique" to transition into "performers" of "the-work-of-Bagouet";
• Problematise the creation of an archive that takes the memory of a subject as its source by historicising and challenging it with other traces and documents.

Since 1993, the collective has debated and adapted practices of transmission and, thereafter, their relationship to Bagouet's work.[3] It is perhaps the work of Bagouet; but at what price, for whom, why, and how? An evolution in attitudes and practices can be found in how the collective has approached revivals. Within the company during the choreographer's lifetime, transmission in the revival of roles was at once permissive, relatively informal, and intuitive. During the period prior to 2000, les carnets bagouet's revivals were produced under the close supervision of two project leads and former performers, and transmission was made more restrictive: through fear of taking Bagouet's vacant place, these revivals examined legitimacy, possibilities, and limitations on modifications. From 2000 onwards, each project lead signalled an explicit commitment to interpretation. New productions and distributions of the choreographic material then became possible, and even new titles for works "based on" the original pieces appeared. Progressively, the concept of rereading, supporting dissemination, and citing the work were asserted. This approach gradually led the collective to no longer think in terms of reconstructing or reassembling the repertoire, but more of deconstructing and disassembling so that new readings of the work could emerge.

Bringing to life in this sense supposes grappling with the work, and, in an instance of double capture, being caught within it, expressing with it, so that other possible meanings are released. Moreover, this situation opened up the possibility of debating the work's future. The Bagouet dancers thus engaged in a political ecology of memorial work that held the potential to establish an alternative model. By developing a culture of reflection and individual and collective responsibility, possibilities for each artist to take charge of the future of works have opened up. In doing so, les carnets bagouet have cultivated the quality of the "irreplaceable" in the sense defined by philosopher Cynthia Fleury, not as guaranteeing a possession or a power, but reaffirming the irreplaceable character of each individual as the first protection of democracy (Fleury 2015: 11–13). Since "we are not replaceable" (Fleury 2015: 11), each protagonist cannot replace another without modifying the project's direction. This does not equate to dissolving the concept of the work (the umpteenth version of the same ballet), nor recycling the materials from it, or even reifying it as a fixed object. The idea of "work-as-starting-point" remains, but it is important to consider these shifts as a different framework of experience each time. This idea is necessary so that the poetic power of a work, which is not flexible but rather plastic, emerges.

Across this significantly debated professional practice (for who, why, and how to pass on forms of knowledge?), transmission takes the form of an evolutive life in a social ecology that belongs to collectives and the contexts to which it is addressed or who wish for it. The work constitutes the very place of debate. les carnets bagouet's experience thus testifies to the possibility of a shift away from a so-called loyalist tradition marked by a desire for fidelity to a creator's work and towards a form of tradition that is more anonymous, sporadic, and discontinuous and understands how to disseminate its knowledge and ethos within a far larger community. These are the necessarily complex traits of transmission, or of this ever-unfinished project that can only be brought to life from the desires and strengths of the dancers who grapple with it to reinvent it.

Notes

1 Here, I am respecting the collective's rejection of capitalising its name.
2 Legrand mentioned this at the seminar "États des lieux", which was held in Magrin (20–25 October 2003). A transcription of these debates can be found in Fonds Carnets Bagouet at the Centre National de la Danse.
3 For a study of "les carnets bagouet" and their archives, dating from 1993 to 2007, see Launay 2007.

Three Key References

Franko, Mark (2017). *The Oxford Handbook of Dance and Reenactment*. Oxford: Oxford University Press.

Launay, Isabelle (2017). *Les danses d'après I. Poétiques et politiques des répertoires.* Pantin: CND.

Launay, Isabelle (ed) (2007). *Les Carnets bagouet, la passe d'une œuvre.* Dijon: Les Solitaires Intempestifs.

32

MUSIC

Carmen Pardo Salgado and Makis Solomos

In this entry, we briefly sketch out how music can be considered within the Guattarian model of the three ecologies that combine to form an ecosophy.

Considering the first of these ecologies, the environmental one, allows us to reassess debates on the relationship between music and nature within a broader context. This might include questioning the boundary between musical sounds and sounds produced by the environment or animals (in particular, birdsong, which has always inspired musicians, but also insect song in more recent music). The well-known work of Steven Feld (1982) has shown the non-dissociability of music and the natural environment in the societies he studied. Another key theme in current artistic practice relating to environmental ecology is connected to the notion of *place*, which can be approached in multiple ways (see Berque 1987; Norman 2012; Esclapez 2014). This theme covers a vast field from artistic projects using *field recording* to site-specific practices (such as the "bell concerts" that Llorenç Barber has staged in several cities: see Claver 2010), via soundscape compositions. It would be impossible to establish even a simple list of musical and sonic questions related to environmental ecology. Such a list would be endless. From Xenakis's composing with cellular automata to model fluid turbulence (Solomos 2005) to ecologist artivists recording the sounds of sea ice melting, as a result of climate change, or listening out for the Antarctic's "ice voices" (see Hince et al. 2015), the ground to cover is truly vast.

Second, considering the mind in ecological terms allows us to revisit fundamental questions about music and develop them in new contexts. An example to cite would be the classic question of the "effect" of music or the "power" of sound, which has been discussed from the ancient debate on the ethos of modes to current theories on musical emotion (see Juslin and Sloboda 2011). Revisiting this question emphasises that it is preferable to speak of empowerment in order

DOI: 10.4324/9781003455523-33

to counter positioning the listener as a passive entity and to consider the relationship between music, listener, and milieu. One could also study the many forms of experiences of subjectivation[1] from an ecological perspective. Naturally, at the heart of this ecology is the question of listening. The ecological turn in art is largely characterised by the development of an "art of listening" – and not only in music or sound art. Another important debate, which begun recently in the music and musicology community, is about the use of sound for the purposes of violence (Papaeti 2013; Volcler 2011). Finally, mental ecology invites us to grapple with the question of the body, which, in the past (thinking particularly about classical music), music has excessively "schematised" by reducing it to the notion of "gesture" (an abstract gesture, only visible on the score).

The third ecology, social, is no less important. Does the "environment" of the sound and the subject-listener not belong as much to society as to the physical space in which the sound is deployed? Are musical emotions not also socially structured? Challenging the theory of music's autonomy, Hildegard Westerkamp cites music anthropologist Allan P. Merriam:

The core of assumptions in Western aesthetics concerns the *attribution of emotion-producing qualities to music conceived strictly as sound.* By this is meant that we in Western culture, being able to abstract music, and regard it as an objective entity, credit sound itself with the ability to move the emotions.

(Merriam cited in Westerkamp 1988: 71)

Westerkamp adds: "in other words, the Western aesthetic separates the experience of music from its social context. When one is moved by the music in that sense, one is moved *internally*, privately, as an individual" (1988: 71). Conversely, the ecological approach, contrary to many sociological approaches, suggests that the environment that makes up society is also a veritable "milieu" that is permanently interacting with the sound and the subject. It thus follows that we can re-examine the relationship between music and society. By avoiding the trap of reflection theories and reconsidering this relationship within the autonomy–heteronomy dialectic proposed by Adorno (1982), we can no longer suppose that everything takes place within music itself. Regarding current artistic practices, it will be important to broaden the ecological debate (in the narrow meaning of environmental ecology) to include social questions by, for example, addressing current debates about collective artistic action or cultural and natural commons.

Turning our attention to the notion of an ecosophical approach that combines these three ecologies, we propose acting and participating in musical practices from a non-anthropocentric perspective. This proposal supposes a questioning of the way that human beings position themselves as the regulating centre of musical experience. A non-anthropocentric perspective shifts the human being from the central position to other mobile locations. It thus recognises that other organic and non-organic beings could become as much centres as human beings and, like them, establish relationships where the positions would presumably not be fixed. This is what John Cage – a musician who carried out the most radical criticism of anthropocentrism in music – describes as an attitude that is not "deaf and blind to the world around [him]" (Cage 1976: 140), "a state of interpenetration and non-obstruction" (Cage 1973: 38).

In systematic terms, imagining music from a non-anthropocentric perspective implies:

1 Speaking of music in a plural sense, and not as a singular music. Music is not a universal construct created according to national, ethnical, gender, or colonial considerations: music is always a multifaceted manifestation tied to its milieu.[2]

2 Not considering forms of music as complete objects that can be examined under the microscope by subjects who are also conceived as complete. Forms of music and subjects are always in transition, in a state of being affected and affecting change. These affections are understood as movements of intensity that do not always enter into the predefined space of representation. In that sense, speaking of affect, once again, poses the question of a musical perception that necessarily goes beyond emotion theories centred on the subject.[3]

3 Acknowledging the sonorous and non-sonorous milieus that make up music. This means actively recognising the relationship between what is called art and what is called nature and life.[4]

4 Referring to musical practices and not only musical works (see Pardo 2016a). Musical practices are forms of music in action, multiple possibilities to say we, they, and so on.

5 Understanding that forms of music contribute to the production of sonorous and existential territories[5] that, whilst acknowledging the ecological turn in art, are characterised by decentred types of listening and sonorous production with regard to the sonorous and musical imaginary of *Homo economicus* (see Barbanti 2016; Paparrigopoulos 2016a; Solomos 2016).

6 Recognising that, sometimes, there will no longer be a perspective, not even a non-anthropocentric one, and that we feel our way through sound.

Notes

1 The notion of "subjectivation", which refers to a process, is usefully substituted here with the classic question of subjectivity (see Guattari 2013: 275–83; Foucault 1984: Introduction).

2 On the notion of multiplicity in music, see Daniel Charles (1998: 118–27).

3 From onomatopoeia in *Zang Tumb Tumb* by Marinetti (1914) to *Apocalypse Will Blossom* by Charlemagne Palestine (2007), via Arnold Schoenberg's *Pierrot lunaire* (1912), Xenakis's *Pithoprakta* (1955–56), or John Cage's *Telephones and Birds* (1977), all open up the range of ways to create affects.
4 As the composer Pascale Criton explains about her music: "It is rather a system of unstable behaviours, able to enter in relation with a changing environment, which does not reproduce itself according to a constant law" (Pardo 1999: 61).
5 "There is a territory precisely when milieu components cease to be directional, becoming dimensional instead, when they cease to be functional to become expressive. There is a territory when the rhythm has expressiveness. What defines the territory is the emergence of matters of expression (qualities)" (Deleuze and Guattari 1987: 315).

Three Key References

Charles, Daniel (1998). *Musiques nomades*. Paris: Kimé.
Feld, Steven (1982). *Sound and Sentiment. Birds, Weeping, Poetics, and Song in Kaluli Expression*. Philadelphia: University of Pennsylvania Press.
Westerkamp, Hildegard (1988). *Listening and Soundmaking: A Study of Music-As-Environment*, MA dissertation. Simon Fraser University, Canada.

33

MUSICAL PERFORMANCE AND WET MARKETS

Pavlos Antoniadis

The negative impact of the COVID-19 pandemic on live performances reached levels not seen since the Second World War (Ministère de la culture 2020). Revenue from concerts collapsed and streaming services, which amount to pure exploitation for performers, dominated the market, reinforcing trends that had already taken root before the crisis (World Economic Forum 2020). Wet markets of live animals, the epicentre of twenty-first-century pandemics, are closely related to music's "wet markets".

What do we mean by "wet"? The term is meant literally in the case of live animal markets where a considerable amount of liquid circulates. Often these markets sell seafood, so the floor is constantly wet due to melted ice or the cleaning of food stands. Wet also refers to the flow of blood from slaughtered animals and the transmission of viruses between species. It is precisely this proximity between humans and animals, a non-regulated and disproportionate proximity that produced the ecological crisis, which threatens to transform "interanimality" and "intercorporeality", concepts borrowed from Merleau-Ponty (Boccalli 2019), as the foundation of the musical act. The physical coexistence between musicians and spectators – which constitutes a "wet market" of desires and products from the musical body, generated emotions, and sometimes blood, sweat, and tears – becomes available for "slaughter", to be put up for sale in a market that is also wet in its own way. The market of streaming digital data connects to a network of brains, and together they form the "wetware" and "netware" of new information and communications technologies, notions that are complementary to the distinction between software and hardware within the framework of "cognitive capitalism" (Moulier-Boutang 2012).

The in-person economy in the arts faces a double-edged sword: at once, there is a crisis of concerts and a crisis due to the dominance of streaming,

DOI: 10.4324/9781003455523-34

which have provoked a multitude of creative responses on an artistic and technical level, feeding speculation on the future of music. These responses range from impromptu solidarity concerts on balconies to virtual rooms at the biggest festivals; from the anarchist proliferation of concerts by precarious musicians in the most intimate spaces to the monopoly of the video conferencing platform Zoom, which is becoming the new tapestry of our online communications; from the fight against latency in sound transmission to researching new technologies of musical interaction. In turn, many questions are raised: what will be the future of performance, particularly musical performances within the context of a widespread digital mediation of interactions between performers and the public? Can the culture of simulating physical and social interactions in a virtual space replace the experience of reality? Can we really distinguish between physical distancing and social distancing? Beyond investing in sound and image quality for remote performances as part of virtual concerts, what brings about the potential integration of data on physical interactions (e.g., movement or haptics) into the Internet of Things or the virtual and augmented reality of the concert? What becomes of musical performance's value in terms of remuneration of online concerts and copyright?

In this entry, I will simply sketch out some brief initial and provisional remarks, which are themselves somewhat "moistened" by empathy with regard to what is seemly being lost as well as by an impatience for a new Guattarian ecology (Guattari 1989) that bridges real and virtual, symbolic and digital, cognitive and corporeal, public and private. As a pianist and musicologist who works with interactive technologies, my reflection is limited to the domain of so-called contemporary music. Other musical genres are certainly more wet, more vibrant and participatory than contemporary music. Yet, the hybrid nature, particularly between ritual and symbolic abstraction, of performing so-called art music renders it appropriate to reflect on the subject matter at hand. In some respects, the music body's transformation in art music has always been technologically mediated; technology has always been its ontology.

This present reflection attempts to go beyond the shock of COVID-19 by contextualising current developments over a longer time period that includes the following series of events: the biopolitical origins of musical performance that play out in the formalisation of disciplinary techniques relating to:

- The body, in music training institutions, in the abstraction/symbolisation that represents musical writing, and in the hierarchy between composer and performer (Lessing 2014);
- The history of musical technology as a history of "mutilation" and "augmentation" of the technical material and agents (see Keislar 2009 drawing on Marshall McLuhan);
- The recording of musical experiences by cognitive capitalism's Web 2.0 devices with a corresponding transformation in the political economy of

performances towards immaterial works, an intangible heritage of innovation and valorisation (Moulier-Boutang 2012);

- The appropriation of private data (biometric performance data) in the near future within the framework of surveillance capitalism (Zuboff 2019) where the modification of human behaviour would be led by markets of wet data collected by "data barons" (Neidich 2013: 15);
- The growing musicalisation of information and communications technologies themselves in the sense of a virtuosity and performativity demanded by the user on the level of interfaces, a musicalisation that shaped the development of these technologies, for example with the graphical user interface of computers, and which is amplified in research on virtual or augmented reality that integrates haptic and biometric interactions (Buxton 2008, cited by Holland et al. 2013: 3).

Moreover, the shock of COVID-19 is but a catalyst for diachronic processes of abstraction and absorption of musical performance within the current apparatuses of cognitive capitalism. By "mutilating" physical coexistence and "augmenting" its digital liquefaction, the current situation follows a constant line of transforming musical performance: from praxis to symbols, recorded physical energies, and digital data. This line of symbolisation, its technical nature in the sense of Bernard Stiegler (1994), can be considered in musical performance as a "dehiscence" (Nancy 2021); not an opposition to nature but a bifurcation in the organic nature of the musical performance, creating a relationship to itself. According to this concept, the body forms a continual spectrum between forms of reality and the virtual world, whether in terms of the slaughtered animality of live performance or in terms of diffused brains streamed in a network.

However, beyond hosting *apéros* over Zoom and other empty rituals of physical distancing, the most urgent question is a socio-political one. Following David Graeber's criticism (2008) of theories of post-operaismo and cognitive capitalism, we must not lose sight of the fact that no new materiality – situated between reality and the virtual world, human intelligence and artificial intelligence (Madlener 2022), the body and digital performance data, "cultural plasticity and neural plasticity" (Neidich 2013: 10) – will remedy Karl Marx's observation that the world is not made up of a collection of discrete objects to buy and sell, but social actions and processes. The value of art, including that of art music performance, is never produced by adapting to music industry circuits, streaming platforms, festivals, artistic directors, the media, and technological transformations accelerated by COVID-19. It is produced by the courage of artists to create post-capitalist enclaves where:

> it [is] possible to experiment with forms of work, exchange, and production radically different from those promoted by capital. While they are not always self-consciously revolutionary, artistic circles have had a persistent

tendency to overlap with revolutionary circles; presumably, precisely because these have been spaces where people can experiment with radically different, less alienated forms of life. The fact that all this is made possible by money percolating downwards from finance capital does not make such spaces "ultimately" a product of capitalism any more than the fact a privately owned factory uses state supplied and regulated utilities and postal services, relies on police to protect its property and courts to enforce its contracts, makes the cars they turn out "ultimately" products of socialism. Total systems don't really exist, they're just stories we tell ourselves, and the fact that capital is dominant now does not mean that it will always be.

(Graeber 2008: 11)

Marginal but porous in their relationship to systemic structures, these self-managed enclaves remain by definition wet.

In memory of David Graeber who passed away on 2 September 2020 in Venice.

Three Key References

Guattari, Félix (1989). *Les Trois Écologies*. Paris: Galilée.
Madlener, Frank (2022). "Intelligence artificielle et imaginaire artistique". *Analyse Opinion Critique*. URL: https://aoc.media/opinion/2022/11/30/intelligence-artificielle-et-imaginaire-artistique/ (accessed 1 December 2022).
Stiegler, Bernard (1994). *La Technique et le Temps 1*. Paris: Galilée.

34

PERFORMANCE

Hélène Singer

Performance revolutionises the relationship between spectators and works of art by forming unprecedented body-to-body connections with artists. It is the *mise en scène* of an intention, or rather the "*mise en tension*" (in Latin, "*intentio*" means "tension"), to reach, to deeply move the spectator. This affect leads to reflection and to the questions raised by the artist, which are often political in nature: social context, feminism, ecology. Performance is thus a sensory experience of the political in which the artist displaces conventions and interrogates our environment and our ways of living in the world. Through physical action, its political engagement is tangible. This sensory engagement separates physical performance from a simple "engaged" artistic action, making it a potential ecosophic practice. Whilst not all performances can be considered as such, this medium, which places the body at the centre of the work, poses de facto (or *in vivo*) the question of living beings' presence in the world.

Artistic Action, and Environmental and Social Engagement

Artists were raising ecological questions long before they started receiving coverage in the media. In 1982, artist-pedagogue Joseph Beuys reaffirmed his ecological engagement by beginning his planting of *7000 Oaks* in Kassel at documenta 7. For Beuys, ecology can only be approached through a desire to modify the social order. Beuys states that engaging in this collective action was "for my enterprise of regenerating the life of humankind within the body of society and to prepare a positive future in this context" (quoted in Demarco 1982: 46). In 2007, British artists Heather Ackroyd and Dan Harvey collected acorns from these trees to replant them elsewhere, thereby transplanting Beuys's original project onto the contemporary ecological field, which is

DOI: 10.4324/9781003455523-35

addressing the exploitation of natural resources. This project, entitled *Beuys' Acorns*, contributes to the continuity of the "social sculpture", a phrase used by Beuys to describe an expanded field of art that is interdisciplinary, engaged, and participatory.

Ecological art seeks to leave a strong visual impression by staging how harmful our power over nature can be. The Argentinian Nicolás García Uriburu worked on *7000 Oaks* after collaborating with Beuys the year before on a project that turned the Rhine River bright green with fluorescein to condemn water pollution. Uriburu, an activist ecologist and close collaborator with Greenpeace, first carried out this process in Venice's Grand Canal during the 1968 Biennale, and he would subsequently repeat it numerous times, including in the East River in New York and the Seine in Paris. In contrast, other actions that seek to condemn damage to the environment reflect more closely those of organisations or politically engaged citizens, which raises the question of their artistic nature.

Conversely, the aesthetic question is at the centre of ecological engagement by raising awareness of our responsibility towards preserving the beauty of the natural world. The feeling of belonging to a common world was originally formed through experiencing this beauty. Today, the feeling of common goods exists due to the damage to the ecosystem. Some artists propose physically exploring our relationship to the natural world through an "original" relationship with nature. In *A Walk to Pikes Peak* (2012), Harrell Fletcher and Eric Steen invited people to trek in the mountains around Colorado Springs over the course of three days to create the experience of a common work in which each participant becomes an artist by studying the environment. By "democratizing the role of the artist through [Fletcher's] socially engaging publicly collaborative projects and installations" (Goldberg 2018: 85), environmental ecology and social ecology overlap. This reflects an ecosophic concept in which our social responsibility cannot be disassociated from the unique experience of our state in the world. We inhabit it, poeticise it to invent new forms that are liberated from the pressure to produce at any cost.

Performance: Fusion and Dissolution of the Hierarchy of Living Beings

Other performers put their physicality to the test in their work. Environmental questions are often embodied through a symbolic dilution of their bodies into the natural elements, particularly water, as if it dissolves the hierarchy of living beings and puts the artist on the same rung as the world's other elements.

In the project and series of photographs *Pink Depression – L'Eau mourante* (1982), the German artist Barbara Leisgen floats Ophelia-like in a river. With her head submerged, she holds her breath to protest against water pollution. A decade earlier, the Cuban artist Ana Mendieta similarly floated nude in a

Mexican stream (*Creek*, 1974), yet with very different artistic aims. There is no strictly political significance here; rather it evokes a conceptualisation of the living being: the fusion of the body with the elements represents a reconciliation with nature. Mysticism and primitivism combine in Mendieta's mental summoning of the natural elements and pagan gods: "I am connected with the Goddess of sweet water" (quoted in Montano 2011: 28). This shift from rational to irrational sensory experience can also be found in the work of Japanese artist Yayoi Kusama. Her concept of "self-obliteration", understood as a dissolution of the self into the environment, was born out of a visual hallucination. At the age of ten, Kusama saw the flowers on a tablecloth multiply on the walls and on her body: "I saw the entire room, my entire body, and the entire universe covered with red flowers" (Kusama 2011). Kusama would translate this concept into a performance by painting dots on her naked body and those of others, signifying that the individual is only a dot amongst the millions of dots that make up the universe. How can the individual inscribe their identity in this jumble of living beings? All attempts seem in vain, like Kusamsa's painting of dots in the water of a pond in which she is immerged (scene from the film *Kusama's Self-Obliteration*, 1967). This practice of perception, in which the psyche redefines the living being, links mental and environmental ecology.

Fusion becomes dangerous when it leads to the obliteration of the individual: a fusional social state may alienate the subject and not allow them to think for themselves or to be unique. Drawing on Mendieta's performances in which she wanted to become a natural element (grass, a tree), Paul Ardenne underscores the philosophical and political limitations of humans' quest for obliteration in nature. Denying oneself can lead to a downward slide into anti-humanism: "As if humans must be ashamed of what they have become" (Ardenne 2019: 81). Fusion for Kusama is indeed auto-destruction, but her orgiastic and sexual actions with others, where bodies intermix, shows that the dissolution of each individual can also be a source of the collective celebration of living beings.

Desire and Limitations of "Becoming Savage": A Fantasy Territory outside of the Cultural Domains

Needing to rediscover one's primitive roots, a lack of differentiation between human and non-human animals, a quest for the golden age of nature that was corrupted by humans: we are witnessing the return of the noble savage myth. The natural primitive state, not dominated by man, is a source of inspiration for performers who see it as outside of the cultural domains and liberated from any social obligation. There is perhaps a Western artistic misinterpretation of "savage thinking" that confuses it with savagery. Nothing is less anarchical than theories that serve to order the universe. According to Lévi-Strauss, rituals, far from being an expiatory release, signify a "'micro-adjustment' – the concern to assign every single creature, object or feature to a place within a class" (Lévi-Strauss 1966: 10). In opposition to the hubris and savagery assumed by the Viennese Actionists, who wish to disgorge civilisation of its repression through sadomasochistic and scatological ritual action, other performers seek to re-establish the link with untamed nature in more peaceful ways as a means of rebuilding society. This might be manifested through communicating non-verbally with animals like Beuys in the film *How to Explain Pictures to a Dead Hare* (Galerie Schmela, Düsseldorf, 1965), or Rose Finn-Kelcey and the two living magpies in *The Magpie's Box* (Acme Gallery, London, 1976). As for Jan Fabre, dressed in raw flesh, he assumes the role of living memento mori in the performance "Ich bin ein Skelettmann" (in Pierre Coulibeuf's film *Doctor Fabre Will Cure You*, 2013), thus professing a finitude shared by all living beings. Lastly, Rémi Voche adopts and subverts the codes of sacred rituals linked to nature in the performance *La Bénédiction du raisin*, the name of a traditional Armenian festival, in which he stamps on red grapes covered with a white sheet (Galerie Ceysson & Bénétière, Saint-Etienne, 2018). Through this symbolic persecution which transforms the white sheet into a shroud bloodied with grape juice, the Dionysiac ritual becomes a political action through its connotation of genocide.

When performers follow the scores of artistic rituals that have social rather than religious significance, there is the potential to align with ecosophic thought. Through a sensory and irreverent engagement, they become agitators of consciousness who reinvent a state in the world and an order of society by uncovering the territories where each element of the living world legitimately finds its place.

Three Key References

Ardenne, Paul (2019). *Un art écologique. Création plasticienne et anthropocène.* Lormont: Le Bord de l'Eau.

Goldberg, RoseLee (2018). *Performance Now: Live Art for the Twenty-First Century.* London: Thames & Hudson.

Wood, Catherine (2018). *Performance in Contemporary Art.* London: Tate Publishing.

35

PEST PLANTS

Lorraine Verner

Categorising certain living beings, whether animal or plant, as "pests" has long demonstrated the anthropocentric vision that hangs over the non-human world. Plants continue to be discussed as pests, even if the equivalent term *"nuisible"* in France has been replaced by the circumlocution "susceptible to causing damage" since the introduction of a law in 2016 to "recapture" biodiversity, nature, and landscapes.[1] The idea of "destruction" is maintained in the distinction between domestic and wild species, with the term "pests" being replaced with "non-domestic" species. The rather astonishing call to "recapture" biodiversity in the law's name attests to a competition that implies the control of one species over another. Non-human beings are still considered according to their degree of utility and monetary value. Each living being is judged on its function for humans, who want to impose order on the world, and is subject to the pressures of our societies: overexploitation of resources, pollution, excessive modes of production and consumption, the destruction and alteration of certain ecosystems, and an erosion of biodiversity.

Whilst the relevance of applying the term "pests" to plants has been questioned, this appellation has previously been subject to various meanings based on the worldview that underpins it. It reflects the evocative, subjective terminology used in French for "weeds", species that are said to cause a nuisance: *"méchantes herbes"*, an eighteenth-century term meaning naughty plants or *"mauvaises herbes"* (bad plants) in current vocabulary. The associated semantic field includes: proliferating, invasive, disordered, useless, pathogenic, crossbred, stray, illegal, wild, parasitic, threatening, harmful, aggressive, killers, *deviant miscreants, green cancer, green Ebola, plant plagues, ecological demons, femme fatale* or even *wicked witch*. As Jacques Tassin and Christian Kull

DOI: 10.4324/9781003455523-36

identify, this rhetoric of alarmist and anthropocentric metaphors falls into four main categories: "military, health and disease prevention, nationalistic reflexes, and the cultural foundations of our societies" (2012: 404), which all shed light on the nature of relationships between humans and their environment.

In the process of *generalised domestication*, Ghassan Hage (2017) asserts that the use of pejorative metaphors, including those relating to plants, presents nature through the lens of racist classifications and reflects the dominant mode of inhabiting the world. This process classifies alterity deemed to be threatening, non-exploitable, ungovernable, and undomesticated, thus metaphorically assigning it the status of a *bad plant*. Some plants can be considered as "practical nuisance[s]" (Hage 2017: 46), which Mary Douglas would describe as "despicable object[s]" (1966: 81) or "matter out of place" (1966: 165). Hage adds further: "The classifications 'useful' and 'harmful' are clearly from the perspective of the domesticator" (2017: 107).

Some characteristics attributed to undomesticated vegetation also evoke the figure of the sorceress and reflect the categories of domination that ecofeminists uncover through connecting the exploitation of natural resources and that of women. As Françoise d'Eaubonne (1974) affirms, this is a double appropriation of the earth and the womb. The same relationship to alterity is at stake.

Others, like myself, belong to the same complex weave that is the fabric of life. Within this framework, Gilles Clément's concept of the "third landscape" (2004) assumes the entire scope of its position, favouring biodiversity, biological forms of resistance, and alterity on the outskirts of domesticated nature. Analogously, this concept has a significant ethico-political dimension: alterity and diversity are considered as necessary for life, not only in biological (biodiversity) terms, but also in social, cultural, and psychological terms (Barbanti et al. 2013: 319).

Different artists' observations on spontaneous vegetation, its biological and genetic resistance to spatial hierarchies and fixed urban and rural planning, can teach us to think beyond the distinction between humans and non-humans and towards a co-habitation developed through interspecies diplomacies and mutual agencies. The underlying question of these artists' works seeks to understand how living vegetation, in its diversity and materiality, can lead to an examination of anthropocentrism's devastating effects and our ways of inhabiting the world we live in. How can art reinvent places of interaction with the living beings around us and within us?

Resisting insignificance and fragility is often employed as an analogy for life and artistic practices, which is represented through a capacity for adaptation, unpredictability, and the mobility of "wild plants", often symbols of liberty and autonomy. In an artist's book, Jakob Gautel (1999) creates "portraits of plants. Almost invisible, but tenacious and full of a passion for living". For the photographer Claude Courtecuisse, who is also an observer of indeterminate spaces, plants in his series *Les petits jardins clos d'herbe errante* (1994) mark the movement of vegetation through which, by analogy, our own certitudes and our relationship to the world can be examined: "A small displacement of things that might be even more justified in the disorder of things" (2005: 68). This subject also interests Lois and Franziska Weinberger who describe their practices that are in perpetual transition and question different hierarchies as a "PLACE / WHERE THE LIVING REVEALS ITSELF ABOVE THE ORDERLY" (1997: 45). For them, the plant world offers humans a mirror to question their milieu and its codes, fears, and rejection of the other. These approaches and practices reflect Isabelle Stengers's remarks: "We are no longer only dealing with a nature to be 'protected' from the damage caused by humans, but also with a nature capable of threatening our modes of thinking and of living for good" (2015: 20).

In 2010, Laurent Cerciat took a series of photos called *Les Rudérales*, which placed in the foreground opportunistic flora around demolition sites in a neighbourhood in the process of regeneration. Observing the diversity and nomadism of seeds documented the different populations he encountered, which consisted of around 30 different cultures living side by side.

Lois Weinberger frequently employs a poetics of the "Ruderal", the free circulation of plants on waste ground, to proclaim the coming of a "Ruderal Society", which would create a fault line within the urban environment (2009: 26–7). Since 1994, he has developed a project of "Portable Gardens" by filling simple plastic carrier bags often used by migrants with earth from the urban wilderness, and in which seeds carried by birds, insects, or the wind can be deposited. Another example is his project "What Is Beyond Plants Is at One with Them" from documenta X in Kassel (1997), where he planted non-native species from south and south-eastern Europe on a disused railway line to cohabit with local vegetation. Weinberger viewed the botanical process of rapidly growing plants as a metaphor for contemporaneous migratory displacements. Through the plant world, he addresses ideas of precarity, hybridisation, and conflictual relationships between the foreigner and the native.

Since 1999, Maria Thereza Alves in *Seeds of Change* has traced an alternative history of colonialisation, migrations, exchanges, and their lasting effects on the landscape through seeds transported in the ballast unloaded by merchant ships. In *The Bank of the Migrating Germplasm* (2016), Leone Contini (2017) sows, harvests, and spreads seeds and plants to explore questions of displacement, hybridisation, and the relationships between native and foreign plants. Some aspects of this work can be situated at the intersection that T. J. Demos locates between post-colonial struggles and environmental concerns in calling for a "decolonizing nature" (2016).

Studying the environment surrounding plants, including those that evade human control, teaches us to reflect on what alterity, hospitality, cohabitation, migration, and responsibility mean. It is necessary for humans to imagine interactions and acts of solidarity with non-humans, our partners in this fragile world. With the mass extinction of living beings underway, it is more urgent than ever. Allowing essential yet sometimes relatively invisible plants to live will allow us to continue living since we depend on this diversity. Today, it is necessary to "cultivate attachments", to use Bruno Latour's expression, with "all those on earth and not just humans" (2017: 106). We can learn from plants, even when they are judged to be a nuisance, non-domestic, or undomesticated. We must be attentive and explore other ways of expanding the "Parliament of things" (Latour 2018), other ways of saying *us* "IN THE COMPLEXITY OF THE UNDETERMINED", which, according to Lois Weinberger (2009: 84), the organic offers us. Building on this idea, Weinberger's installation *Laubreise* (2009) examines the making of a better earth, a better world conceptualised as networks of relations and shared land.

Note

1 See "Loi pour la reconquête de la biodiversité, de la nature et des paysages". Available online: https://www.legifrance.gouv.fr/jorf/id/JORFTEXT000033016237 (accessed 20 January 2022).

Three Key References

Hage, Ghassan (2017). *Is Racism an Environmental Threat?* Bristol: Polity Press.
Tassin, Jacques, and Christian A. Kull (2012). "Pour une autre représentation métaphorique des invasions biologiques". *Natures Sciences Sociétés* 20(4): 404–14.
Weinberger, Lois (2009). *The Mobile Garden*. Bologna: Damiani.

36

PHOTOGRAPHY

Michel Poivert

On the margins of the social phenomenon known as the economy of digital images, artists are experimenting with processes that combine photography's predigital heritage and contemporary art, in which the concept of post-photography is freely discussed. With the economy of attention having been profoundly altered, new ways to practise photography are circulating. What does the term "photography" mean in these conditions? Rather than a practice with the sole purpose of producing images, photography has the potential to interrogate our relationship to perception.

Concept

Influenced primarily by the avant-garde movements that reunite art and craft, a new generation of artists is forming a counter-cultural current that is attentive to threatened forms of expertise and openly critical of technology. The question of the visual's materiality is returning to the centre of debates. Bringing together heritage and innovation, re-establishing a slower temporality, combining art and science with more environmentally friendly processes, and adapting to shared social practices: these new photographic practices are creating an art of transition.

The almost complete disappearance of film photography at the turn of the twenty-first century is the result of a technical and industrial decision. Transferring one part of the photographic process to new technology dissolved physical photographs into an economy of digital images. This transformation has been accompanied by a phenomenon that may seem contradictory: greater historical awareness has brought heritage processes into contemporary practices and found enthusiastic practitioners amongst artists and amateurs alike.

DOI: 10.4324/9781003455523-37

In this context, predigital photography designates a culture offering expertise as an alternative to the new visual standards. As this new approach develops, numerous factors have spilt over from usages normally reserved for visual data. From there, any form of hybridisation becomes possible. Photography is no longer a practice of the past with techniques set in stone, but a repertoire of processes available for aesthetic and social experimentation. We can thus speak of an alternative culture constructed from a photographic praxis and understood as a way of practising and reflecting on photography.

Sensitive Material

Practising photography is thus about understanding how to produce a print, rather than capturing an image. What does it mean to create a photograph during a period marked by an influx and dematerialisation of visual data? Simply that the photograph cannot be reduced to the notion of the image; it must encompass the processes of its materialisation.

What is the difference between an image and a photograph? An image can exist on any material or virtual support, so an image is independent of its materialisation. A photograph, in contrast, is dependent on its processes and, in short, the material on which it is printed: glass, metal, paper, plastic. A photograph can be an image and an image can be a photograph, but not all photographs are images and not all images are photographs. This distinction establishes a new concept for the photograph: its destiny is not necessarily linked to that of images.

The primacy of photographic material over the image produces a relationship to reality that is very different from that of a representation or a recording. The value of materials and the art of applying them establishes this relationship on production. From printing to sculptural installations, post-photography draws on the possible worlds of the photographic and redeploys them in new ways. The result of this approach is that before producing a representation of the world, the photograph is a world in and of itself.

Environment

Photography necessitates a learning process conditioned by principles in which time is a central factor. The inherent temporality of most predigital processes (delay between capturing and the end result due to development and printing time) becomes a paradigm that implements a particular relationship to the world based on a slowing down.

The time required to produce a photograph – the very lack of immediacy in its processes – introduces the notion of scarcity within a culture of abundance. Reducing the influx of images is its first ecological advantage. This frugality, however, should not prevent us from going further.

From its origins, the photograph has been linked with the history of chemistry. The list of its pollutants is long, starting with mercury, which was necessary for developing the daguerreotype. It would be dishonest to present digital images as completely harmless in comparison: the carbon footprint of the digital visual industry is far from neutral. Nevertheless, the principle of photosensitivity at the heart of predigital photography can be made green by reorientating its archaic processes towards an ecological alternative. Nicéphore Niépce's invention was based on the photosensitivity of Bitumen of Judea, an organic resin, and this process can be applied to other common substances, notably chlorophyll, that can capture an imprint of light. Discovered by Mary Somerville (1780–1872), the anthotype is contemporaneous with the first photographs, but the process was abandoned due to difficulties in making its results permanent. Today, anthotypy is practised by artists like Christine Elfman and Léa Habourdin. The researcher Anne-Lou Buzot has dedicated numerous works to the method. In terms of the materials required, the exposure of a sensitive surface can be carried out in broad daylight, with or without a lens (pinhole). A photograph thus can be produced with very few means. It might not be permanent or able to be circulated, but the photograph in this sense is an *activity*, more than a regime of productivity.

The traditional practices being used in contemporary photography (photo povera, slow photo) acquire an aesthetic and ethical dimension in the digital context. Experimenting with and researching light phenomena (e.g., phosphorescence) and photosensitive materials can enrich or contribute to knowledge in the field beyond a culture of mass production and technological exploitation. Whilst they may be marginal, these practices have an educational value: making less images, learning to adapt to another temporality, and exploring renewable resources can oppose thinking within the current industrial model founded on increasing digital files and their storage, and, for printing, the widespread use of pigment-based printers whose ecological impact has not yet been established.

Responsible photography is therefore not an asceticism of image, but an ethical and educational basis for ecosophy in a universe overflowing with images. Moreover, the photographic activity introduces practices of sharing and learning where both the laboratory and the shoot are based on exchange, thus becoming a source of sociability.

Politics

Throughout its history, photography has been credited with two opposing values: as an art accessible to all, it has an emancipatory power, whilst its role in propaganda and surveillance means it also has the power to alienate. Simultaneously as a democratic utopia and a dystopian society of spectacle, photography remains ambivalent in character. However, when it loses its value of use in the sense of a utilitarian end, it is freed from and surpasses its applications. Photography's emancipatory power is found in the areas of creation and education. The predigital dimension also provides a way for photography to critique its own history.

Practising historical photographic methods is to undertake an archaeology of an instrument of power: the mechanisms of domination at work in the body of iconography are revealed through the medium itself. For example, presenting a portrait of an individual from an ethnic minority using nineteenth-century techniques highlights the period's construction of racial stereotypes. For example, Kali Spitzer and Will Wilson use the tintype process based on collodion, a photosensitive substance that darkens skin colour and created the "red skin" image of the indigenous people of the Americas. Returning to the origins of visual archetypes through the images themselves is to highlight their power and the necessity of undoing it without celebrating it as heritage. Reducing traditional processes to heritage would relegate the roots of discrimination to a neutral past of history. In other words, it would be accepting an aesthetic based on its historical value for museums. When contemporary artists employ predigital techniques, it shows these processes still have relevance, thereby challenging their status as a fixed heritage and creating from them a revived language that can be reclaimed. More fundamentally, it makes us aware of the responsibilities of material representation, thereby questioning the material itself and, by reusing it, revealing how a representation is embodied, manipulated as much as visually perceived, in the flesh of the one who binds.

Predigital culture enriches post-photography. This historical recycling affords photography a new place within contemporary art. The photographic processes themselves lead to greater political awareness.

Three Key References

James, Christopher (2015). *The Book of Alternative Photographic Processes*. Boston: Cengage Learning.

Poivert, Michel (2022). *Contre-culture dans la photographie*. Paris: Textuel.

Shore, Robert (2014). *Post-Photography: The Artist with a Camera*. London: Laurence King Publishing.

37

PLACE

Augustin Berque

Place from the Latin *plateia*, meaning broad way, public square.

In abstract terms, a place can be defined as a point within a Cartesian coordinate system, but this abstraction will not concern us here. In concrete terms, a place is *where there is* something or somebody. It is thus related to the existence of that thing or person: if there were nothing, there would be no place and no need to worry about what a place is. Yet when there is a need, as there is now, it means that something is *going on* and, accordingly, that a place is not only a matter of space, but of space–time.

The question of "place" is not new. Both Plato and Aristotle considered the subject. Plato did so in the *Timaeus* in relation to what he called "chora χώρα", and Aristotle in *Physics* (Book IV) in relation to what he called "topos τόπος". Of course, with a matter so apparently banal as place, they also employed these terms on other occasions, as many people did before them since these words were in common use at the time. However, both Plato and Aristotle tried to define the terms rigorously and in accordance with their respective philosophies.

These two examinations resulted in two a priori incompatible conceptions of place, concerning, among other things, the relation between a place and what there is there. For Plato, in the *chora*, there is what he calls "genesis γένεσις": relative Being, becoming Being. Genesis sits in opposition to the *chora*. The latter is compared here to a matrix, a mother (50d2), or a nurse (52d4), but elsewhere with its opposite, an imprint (50c1). How can the *chora* be at the same time one thing and its opposite? Plato eventually refrains from rationally defining what it is. For him, *chora* belongs to "a bastard reasoning" (52b2), "hardly believable" (52b2); all in all, "when seeing it, we dream" (52b3).

DOI: 10.4324/9781003455523-38

The *chora*, that "third and other gender" (48e3), which is neither absolute Being nor relative Being, is removed from the equation.

Aristotle, on the other hand, has a very clear conception of *topos*, as well as of its relation to what is within it, which he calls "pragma πράγμα" (a thing, a matter). A *topos* is like an "immobile vase" (212a15) containing what is there, and precisely limiting it. Hence his definition of place is "the immobile immediate limit of the envelope" (212a20) of the said thing.

What is the essential difference? It is that the Aristotelian *topos* has no ontological relation to the pragma that is there. Indeed, it is immobile, whereas the pragma is not. They can thus be separated, while keeping their respective identities. In contrast, *chora* and genesis are ontologically indissociable since they are mutually the imprint and the matrix of one another.

Moreover, it is not only a matter of ontological, but also logical, difference. Indeed, the relation between *topos* and pragma belongs to an (Aristotelian) logic of the identity of the subject, since the two logical subjects, *topos* and pragma, keep their respective identities even when separated. Whereas in the *chora*/genesis relation, while each term does have an identity (*chora* is not genesis), these two identities are not separable: they belong to each other, and therefore are subsumed under a superior identity, an identity of a "third and other gender", which comprises both. In other words, it is an identity that would at the same time be A (genesis) and non-A (*chora*), thus alien to Aristotelian logic, which does not allow for contradictions (nor does Plato's rationalism, and this is indeed why Plato gave up thinking about that "third and other gender", leaving it to dream).

We have thus, between *chora* and *topos*, an onto/logical (both ontological and logical) disagreement. Let us first say that, over the course of the history of place, at least in the West, it is *topos*, together with Aristotelian logic, that prevails. However, this was a Pyrrhic victory, because while affirming itself onto/logically, this concept of *topos* produced a civilisation that will not rest until it displaces and relocates beings and things: in sum, an atopic, u-topic, place-negating civilisation. Take, for example, Descartes:

> Then I examined with attention what I was, and I saw that I could pretend that I had no body and that the world and the place where I was did not exist [...]. From that I recognized that I was a substance whose essence or nature is only thinking, a substance which had no need of any location and did not depend on any material thing.
>
> *(Discourse on Method, IV, 4)*

Now, if the *cogito* (the modern subject) thinks that it has – putatively – no need of a place in order to exist, it is because it thinks of place as a *topos*, separable from what is there, and, by that fact, that its own identity has nothing to do with that of this *topos*. It is precisely according to the same onto/logical principle that the modern architecture movement reproduced the same "international style" parallelepipeds across the "universal space" of the planet. We are left with the E.T. architecture of our starchitects, which comes down from the stars to land here just as it would land there, with no regard for the place, and produces the "junkspace" that Rem Koolhaas (like so many others) makes merry and ¥€$ with.

Junkspace is but one expression of the general movement of disearthyfication that characterises our civilisation. It is onto/logically linked to the already ongoing Sixth Massive Extinction of life on this planet. To counter this threat, if it is possible, we have to think of place anew, not only as a *topos*, but also as a *chora*, that is as indissociable from what is there, in particular our Being – and reciprocally, we have to think of our Being as indissociable from a place on the Earth.

What does this mean? That, since it does not exist nowhere, a place must be understood also as a milieu (an ambient world); it exists somewhere for a certain being within the ambient world of that being. This implies a problematics of mesology (Uexküll's "*Umweltlehre*"; Watsuji's "*fudoron*" 風土論).

As far as we humans are concerned (since we are the cause of the Sixth Extinction), what mesology shows is that milieus are not only ecological, but they are also eco-techno-symbolical, and so are the places that compose them. What is symbolic about places and their milieus does not come from within Aristotelian logic (since the principle of symbolicity is that A is non-A), nor within the idea of *topos* (since, owing to symbols, what is neither here nor now is, here and now, biologically alive in our synapses). However, at the same time, what is necessarily ecological and thus physical about places indubitably comes from within a *topos*.

A place then is both a *topos* and a *chora*. This notion requires a transmodern way of thinking, that is, overcoming onto/logically the modern paradigm. In other words, we need mesology's mesologic, acknowledging the ambivalent truth of both Husserl's Earth, which does not move (*die Ur-Arche Erde bewegt sich nicht*), and Galileo's Earth, which nevertheless moves (*eppur, si muove*). And this is an urgent matter, because each and every day, the Anthropocene accelerates the Sixth Extinction, which will, of course, also be our own extinction. Indeed, in order to exist, we humans need a place within an eco-techno-symbolically liveable milieu, not nowhere but somewhere on this planet.

Three Key References

Augendre, Marie, Jean-Pierre Llored, and Yann Nussaume (eds) (2018). *La Mésologie, un autre paradigme pour l'Anthropocène? Actes du colloque de Cerisy-la-Salle*. Paris: Hermann.

Berque, Augustin (2019). *Poetics of the Earth: Natural History and Human History*, trans. Anne-Marie Feenberg-Dibon. London: Routledge.

Paquot, Thierry, and Chris Younès (eds) (2012). *Espace et lieu dans la pensée occidentale*. Paris: La Découverte.

38

PLASTIC ARTS

Lorraine Verner

In the face of a familiar yet deteriorating milieu, Glenn A. Albrecht (2011) writes:

> This era could be called the Obscene, not the Anthropocene. I for one, a human, do not wish to be associated with a period in Earth's History where the dominant people in one species wipe out the foundations of life for all other humans and non-humans.

Albrecht's 2019 book *Earth Emotions* further develops this sense of hopelessness caused by devastation to his environment through defining a new terminology of what he calls "eco-emotions". As Jean-Toussaint Desanti points out, "the Latin word *obscenus* comes from the language of fortune tellers: it designates a bad omen, an unpleasant sign" (1983: 128). Jean-François Lyotard, however, questions how we would interpret a world that no longer offers us any signs: "What can we still say if the silence is absolute and outside us?" (2013: 70). We philosophise because we are exposed to the world. He adds further: "there is no absolute silence, precisely because the world is already speaking, even if confusedly [...]. You can transform this world only by listening to it" (2013: 122–3).

In *Silent Spring* (1962), Rachel Carson evokes the spectre of silence felt in a landscape deprived of birdsong, one of the devastating effects of pesticides and pollution. In *The Great Animal Orchestra*, an immersive installation in collaboration with the collective United Visual Artists at the Fondation Cartier's 2016–17 exhibition, Bernie Krause lets us hear and see this slow movement towards an impoverishing of soundscapes, the world's non-human sounds driven out by the "human din" (2015).

DOI: 10.4324/9781003455523-39

With the world reduced to an ob-scene, unseen scenery, devoid of meaning, mute, interpreted solely through the human lens, David Abram encourages us to revive the links we experience with the living world by making ourselves more available to consider and perceive them as part of an ecology of sensory experience. Both in the ecological circumstances of a ravaged earth and within our personal and collective lives, as Abram continues, "clearly, something is terribly missing" (2013: 343). Affected by this lack, we need to regenerate the interrelations between our ways of living and of feeling and the things and beings that solicit us. Moreover, Gilles Clément states the necessity of a paradigm shift that would emerge from new ways of thinking and perceiving reality:

> To say that, from now on, the future will play out based on a new paradigm is not a hypothesis. [...] It is necessary to immerse oneself, [...] rethink one's position in the universe, no longer place oneself above or at the centre of but within and alongside. [...] That is the work of a generation to come for whom life will not be a game of chance and necessities but an arrangement with an expanded complexity of living beings.
>
> *(2014: 26–9)*

How might this crisis in our relationship to other living beings be integrated into the field of plastic arts,[1] through ways of feeling and acting? What connections can be made between a world that we must once again learn to perceive in all its dimensions and the world of art? To underscore the important potential for awareness of the role played in a changing world by plastic artists who are active in this new paradigm, we can cite the German definition of plastic arts, *bildenden Künste* (see also Chateau 1999), in which form, formation, and transformation are all present. The translators of Joseph Beuys's biography outline how this definition nourished the artist's practices:

> Not only does the verb *bilden* mean "to build" or "to shape", but it also means "to educate", "to instruct", "to shape" in the spiritual sense of the term. The word *Bildung* is practically synonymous with cultivation, both individually (development) and collectively (civilisation). Beuys's rationale relies on this network of meaning: history is sculpture because "das Bildende", the formative and creative element that gives its name to the plastic arts, is at work within it.
>
> *("Note du traducteur" in Stachelhaus 1994: 63)*

For Beuys, a work of art was "a plastic creation that not only shapes the physical matter but could also shape the spiritual matter" which "leads to the idea of social *plastic*" (1988: 13). The "social sculpture" or the expanded field of art no longer figured in his work as a practice beyond the world; rather it opened up the real world in all of its complexity.

More recently, SAFI,[2] a collective of plastic artists founded in 2001 by Stéphane Brisset and Dalila Ladjal, invites audiences to participate in sensory experiences and collective physical activities outside the studio: walking, awakening the senses, harvesting, tasting, cooking, drawing, gardening, DIY, and many other ways of "weaving a milieu". The collective's website cites Augustin Berque (1987):

> It is through the senses that we can understand and access things [...] from the point of view of the relationship between human beings and their milieu, [...] ideally, we would be able to think from the perspective of one's life itself, rather than abstracting ourselves from it.[3]

SAFI's practices thus respond to issues by exploring non-human resources and by interacting with the places they encounter. Creating these moments to build links and redeploy ecosystems is also richly instructive for understanding which paths to relationships between human beings are possible.

Through *Wagon Station Encampment*, a site consisting of tiny camping pods and communal areas, and the *Institute of Investigative Living* as part of the artwork *A–Z West* (2000–), the artist Andrea Zittel opens up new possibilities or ways of living without commodities and sparingly using resources from California's High Desert, near Joshua Tree National Park. It is a space of continual experimentation that interweaves objects, spaces, and everyday actions. Zittel outlines the role conferred to art:

> For me, it is a way of making sense of the world and reconciling things. But essentially, my spiritual state is explained by a profound desire to change the way we perceive things, rather than changing physical reality as it is.
>
> *(Zittel and Mast 2016: 120)*

In 1972, the multidisciplinary artist herman de vries wrote "my poetry is the world", a short manifesto in the form of a poem, translated into 64 living and extinct languages and published as an artist's book (2002):

> my poetry is the world / i write it every day / i rewrite it every day / i see it every day / i read it every day / i eat it every day / i sleep it every day / the world is my chance / it changes me every day / my chance is my poetry.

Poetry appears in the real world, to our senses under constantly changing forms. herman de vries's concept of art, as Anne Moeglin-Delcroix highlights, is as close as possible to "presence – of the world to ourselves, of ourselves to the world" (2006: 431). As the artist walks in a contemplative and attentive way to experience nature, he explores, marvels at, traces, brushes against, feels, lingers, samples natural elements as if they are the primary materials for a future work: "it is not so much about representing but documenting: I only cite nature" (see "Positions d'herman de vries" in vries 2001: 30); "letting nature express itself, [...] drawing attention to that expression and [...] limiting artistic activity to protecting and presenting nature" (see "le bois sacré" in vries 2001: 48). Starting from the position of a general loss of knowledge about the plant world, herman de vries's exploration links scientific concepts to aesthetic experience, which provides access to a perception of our milieu in all the diversity of its phenomena.

In some of the practices discussed here, art is not art as an end in and of itself. Art activates our capacity to perceive what we do not see or what we no longer see of the earthly universe we inhabit. It encourages existential forms of learning and models or reinvents our ways of living and of forging high-quality relationships with living beings, others, and ourselves. It is an open and concrete philosophy as herman de vries states: "for me, plastic art contributes to raising awareness" (2001: 46). Krause also calls upon us to make ourselves more available to the signs of other forms of life and to ecosystems:

> Perhaps, though, as we begin to unravel a few omens expressed through biophonies and geophonies, [...] we will need to more fully understand what exists. Natural sounds [...] are the voices we need to heed closely. For they are balanced somewhere between creation and destruction – and we silence them at our own peril.
>
> *(2015: 150–1)*

The practices that fall within the plastic arts seek to open up experimental spaces that raise awareness of complex forms of solidarity between species, individuals, and living spaces by calling on them to observe, feel, name, and welcome them in all their fluidity. There are so many ways, always situated and immersed, of creating community and exposing oneself to how the earth speaks. Perhaps, this is one of the available routes towards what Albrecht describes as the possible passage to the "Symbiocene", which generates other accounts of our era based on symbiosis and mutual assistance all the while marking a rupture with the "Anthropo-Obscene" or "Anthrobscene".

Notes

1 In a narrower definition, the term "plastic arts", derived from the word "plasticise", refers to practices that involve modelling or moulding. In a more general sense but less often, the term "plastic arts" (in French "*arts plastiques*") can be used for "visual arts" in English, a broad category that includes polysensory artistic practices and diverse forms of expression.
2 SAFI is an acronym for Du Sens, de l'Audace, de la Fantaisie et de l'Imagination, meaning senses, audacity, creativity, and imagination.
3 See SAFI's website: https://collectifsafi.com/nous/ (accessed 27 February 2022).

Three Key References

Albrecht, Glenn A. (2019). *Earth Emotions: New Words for a New World*. Ithaca and London: Cornell University Press.

Beuys, Joseph (1988). *Par la présente, je n'appartiens plus à l'art*. Paris: L'Arche Éditeur.

Vries, Herman de (2001). *herman de vries les choses mêmes*. Lyon: Réunion des Musées Nationaux/Musée départemental de Digne.

39

RECYCLING

Gala Hernández López

To speak of recycling in the arts supposes taking as a starting point the idea of a theoretically unavoidable "becoming waste", but that approach can be avoided. Instead, the focus here will be on how waste's existence is conferred a new functionality, a new meaning, a second life through recycling. By reintroducing obsolete or broken materials into the manufacturing cycle and preparing them to be "reborn", a process of valorising waste is enacted. Above all, these materials are given specific, sustained, and loving attention. The symbolic figure ideal for illustrating certain artists' interest in waste is Benjamin's "*Lumpensammler*" or a Baudelairean rag-and-bone man-poet who revolts against society at the height of capitalism. After having gathered "like a miser guarding a treasure" the waste of the big city, "everything it has scorned", collecting "the annals of intemperance", he makes judicious choices to transform it into "useful or gratifying objects" (Baudelaire cited in Benjamin 2003: 48).

Recycling in the arts thus implies first reorientating attention towards objects that are normally disdained and rejected by the masses. An act of decontextualisation, a change in environment, a reshaping then follows. Waste is saved from being forgotten or disappearing, new meanings emerge, and the rags of history are spoken about differently: unveiling divergences or secret connections between waste items reveals what is unacknowledged, the zeitgeist. By shattering the present into a constellation of materials, forms, words, textures, or images, art can become the medium to critique an era characterised by its semiotic plurality. The artist's goal is thus to bring out the voices of silent signs, give them a new place, rehabilitate them, "allow [them], in the only way possible, to come into their own: by making use of them" (Benjamin 2002a: 460). The artist subscribes to an ecological function: by inventing a new object

DOI: 10.4324/9781003455523-40

or new work, the creative act results in highlighting the invisible or ignored relationships between already existing phenomena and entities.

The history of recycling in art is directly linked with that of appropriation: collages, readymades, Dadaist reinterpretations, pop art, picture generation. As the historical development of art movements shows, the artistic act of appropriation has been constantly repeated throughout the twentieth and twenty-first centuries, often with the aim of criticising the materialism rampant in a society of consumerism and capitalist waste or the dearly held concepts of modernity, such as work, creativity, and the author as artist. Examples of critical reappropriations exist, like found footage or recycled cinema which are built on ruins and against the hegemony of dead, formatted, or inoffensive images. The works of the masters of experimental cinema in the 1960s and 1970s, such as Bruce Conner, Ken Jacobs, Paul Sharits, Malcolm Le Grice, and Morgan Fisher, are most often selected by cinema historiography. The response in France was *détournement*, a technique developed by the situationists, which draws on images from already existing films, like the lettrist Maurice Lemaître's *The Great Train of History* (1978), and Jean-Luc Godard's *Histoire(s) du Cinéma* (1988–98). More recently, the works of filmmakers, such as Martin Arnold, Peter Tscherkassky, Gustav Deutsch, Bill Morrison, Matthias Müller, Harun Farocki, Peter Forgács, Yervant Gianikian, and Angela Ricci Lucchi, have brought new variations and dimensions to the virtually infinite forms of "second-hand cinema" (Blümlinger 2013), demonstrating that all visual memory can be reappropriated and reinvented, politically and poetically.

Global digitalisation and big data have led to an explosion in the production and circulation of information and images, and thus logically to a rise in "lumpen-data", digital and physical waste, the refuse of the "globalised iconomy" (Szendy 2019). Art is called upon to profoundly redefine the self to confront the contemporary crisis of representation that accompanies the climate crisis. Reconsidering the socially dominant and institutionalised individual and collective imaginaries to respond to the chaos of global culture and the challenges posed by the Anthropocene is an ongoing task for art. It is only by recycling or rather upcycling materials that art can dis-assemble and then make perceptible the modes of manufacturing standardised representation that underpins the capitalist macrostructure.

Essential to these emancipatory artistic practices are: practices that advocate for a sustainability of the gaze; an ecology of media, images, and words that is capable of producing aesth-ethical as well as poetic and political connections, capable of historicising the present and opposing the unrelenting rationality of algorithms with human intelligence and feelings; the multiplicity of beings; and the contingency of the world. An authentic, technical, sociocultural transformation needs to take place. This transformation would reorientate the uncontrolled production of images and objects, their commodification and their privatisation, towards a public processual recycling, a creolisation of content, towards a common reappropriation that is constant, horizontal, and democratised beyond commercial circuits. On the internet, this fluid, dynamic, and reactive milieu could be illustrated by a Möbius strip, the universal symbol of recycling: a global cycle, open to all, in which "poor images" (Steyerl 2012) continually circulate between prosumers who use, reinvent, reference, edit, cut, recreate, remix, and finally share them with others. By reinserting these images into the collective circuit, other users can download and use them, which infinitely extends the ecology of the imaginary and forms. Digital reproducibility is accompanied by a viral appropriability, evidenced in memes and generalised practices of online remixing, which cinema and digital art are drawing upon. Increasing numbers of audio-visual works appropriate and recycle, for example, user-generated content and amateur videos taken from the internet: Natalie Bookchin's *Mass Ornament* (2009) and *Testament* (2009–17), Peter Snowdon's *The Uprising* (2013), Philip Scheffner's *Havarie* (2016), Grégoire Beil's *Roman National* (2018), Zhu Shengze's *Present.Perfect* (2019), Kenneth Goldsmith's "uncreative writing", the work of Steve Giasson and Franck Leibovici, Lina Majdalanie and Rabih Mroué's *33 tours et quelques seconds*, and the art of Chris Alexander, Cory Arcangel, Jean-Baptiste Michel, and Penelope Umbrico. These examples highlight possible avenues to explore in this reformulation and critical recreation of cyberspace by the arts.

In this post-media era, reusing images and signs, reindividualising communicational digital technologies – devices that produce subjectivities – is the route towards a collective and creative singularisation. This cannot be conceptualised without rebuilding an alternative and deterritorialised cyberspace where "a phantasmatic economy that is deployed in a random form" would prevail (Guattari 2000: 57). The act of reappropriating and sharing becomes a vector of subjectivation: our "computer-aided subjectivities" (Guattari 2000: 38) are reinvented, reformulating the arrangements of enunciation and salvaging their political power. The obstacles to this ecological utopia are nevertheless numerous. Amongst other barriers, the legal obstacles include image rights, copyright, and intellectual property rights, which block reproducibility by considering common goods of collective interest as private commodities. It is for us to plot against these challenges together.

Three Key References

Chabert, Garance, and Aurélien Mole (2018). *Les artistes iconographes*. Dijon: Les Presses du Réel.

Gunkel, David J. (2015). *Of Remixology. Ethics and aesthetics after remix*. Cambridge, MA: MIT Press.

Marczewska, Kaja (2018). *This is not a copy. Writing at the iterative turn*. London: Bloomsbury.

40

SITE SPECIFICITY

Lorraine Verner

In "Notes on Sculpture", one of the founding texts on *in situ* or site-specific works, Robert Morris demonstrates that within American minimal art from the 1960s onwards, "the situation [in which we experience art] is now more complex and expanded" (1966: 23). According to him, the understanding of a given space, of one's own body and those of other visitors, comes from recognising the sculpted work as a gestalt object, as a form perceptible as a whole. Positioned at the centre of this constructed situation, "the object is but one of the terms in the newer aesthetic" (1966: 21), a driving force that connects kinaesthetic experiences. The work is not determined by the power of the eye alone, but it is dependent on the whole body, the presence of receivers. As we walk between the elements of the work, mobility is integrated into the system. The situated perception of the receiver structures the work. From this point of view, as René Payant explains, it is rather "more from the order of *Gestaltung* than *Gestalt*, and its equilibrium resides in a movement that perpetually displaces its centre, that is to say its centralisations according to the viewers" (1981: 131). It is a question of integration into a specific context, of maintaining contiguous and reciprocal relations with the place where, what Morris calls, the notion of "the experienced variable" (1966: 23) can emerge. On this subject, Miwon Kwon alludes to "a phenomenological [...] understanding of the site" (2002: 3).

The concept of the work thus shifts from an object in and of itself to how it is situated within a real space, at a specific point in time. Robert Barry takes this approach with his wire projects: each is "made to suit the place in which it was installed. They cannot be moved without being destroyed" (Rose 1969: 23). Similarly, for Vito Acconci,

DOI: 10.4324/9781003455523-41

the idea of works in situ – to create something specially designed for a specific place – was very important in the eyes of my generation. [...] It is a way of saying that the work means something here and now. Elsewhere and in other circumstances, it means nothing or means something very different. That is what I really like about *site-specificity*.

(Gintz 1992: 12)

The pioneering site-specific works of the 1960s and 1970s marked a rupture with the notion of pieces of art being transportable and marketable objects.

In turn, Jean-François Lyotard (1981) and Jean-Marc Poinsot attribute the origins of the term to the artist Daniel Buren and his *works in situ*: "the notion 'in situ' indicates an organic and explicit link between the chosen elements and their situation. Daniel Buren developed the concept based on this principle" (Poinsot 1989: 98). To quote the artist himself: "the phrase 'work in situ' can be understood as 'transformation of the host space'. [...] That transformation can be undertaken by the place, against the place, or in osmosis with the place" (cited in Poinsot 1986: 91). Buren used the term *"in situ"* for the first time in 1971 in reference to his controversial and short-lived contribution to the Guggenheim International Exhibition. His works explore the site that acknowledges, what Poinsot calls, the circumstances of its setting (1989). *In situ* works propose a critique of the institutional conventions of exhibitions (Buren 1973b), artistic spaces (Michael Asher, Marcel Broodthaers), and the status of art works (Hans Haacke).

Addressing site specificity, Poinsot highlights, in particular following Richard Serra, Walter De Maria, and Robert Smithson, that many artists and critics refer to the site in terms of its geographic localisation. Yet, this concept is too limited. Artists have developed strategies outside of the traditional places reserved for art through a desire to distance themselves from these spaces. For example, Smithson has created site-specific works that extend the reach of his projects to urban and non-urban public spaces. When he exhibited his work in galleries, the relationship between the exterior and interior (non-site) spaces was presented in a way that interrogated the concept and system of exhibitions. In response to Serra's controversial work *Tilted Arc*, Douglas Crimp affirms that "the true specificity of the site [...] is always a political specificity" (1986: 55). Serra takes this further: "there is no neutral site. Every context has its frame and its ideological overtones" (1980: 168). The meaning of site specificity expands from Poinsot's limited notion to refer to a concept that is more than simply the site itself.

Gradually attention has shifted away from preoccupations about the physical conditions of a site to include the art work's discursive and contextual conditions. Peter Weibel introduced the term *Kontext Kunst* (context art) to describe the practices at the beginning of the 1990s that were centred around their social and ideological setting. For Weibel, the difference between these artistic generations, from early *in situ* projects to considering the context as a social construct, resides in the fact that

> the "critical boundaries" have been pushed back and extended, [...] artists have begun resolutely to take part in other discourses (ecology, ethnology, architecture and politics). [...] It is no longer solely about the critique of art's systems but the critique of reality and the analysis and creation of social process. [...] The interaction between artist and social situation, between art and extra-artistic context has led to a new form of art, where both come together.
>
> *(Weibel 1994: 51)*

Hal Foster defines sites as

> a series of shifts in the *siting* of art: from the surface of the medium to the space of the museum, from institutional framers to discursive networks, to the point where many artists and critics treat conditions like desire or disease, AIDS or homeless, as sites for art.
>
> *(1996: 184)*

For John Lindell, a member of the artistic collective Gran Fury, homosexuality becomes the site of his practices:

> In terms of my own work, homosexual desire is a site and the gay world at large is a site. Again I'm trying to loosen up the notion of a physical site: a site may be a group of people, a community.
>
> *(1994: 18)*

Within this shift from the community as site or from a short-lived intervention at a site to the community in and of itself, numerous socially engaged collaborative participatory projects (Superflex, Dialogue, Ala Plastica, Park Fiction) have been established since the 1990s. Grant H. Kester is interested in open, non-prescriptive, and dialogical interactions between participants and the chosen site, which he describes as "the habitus of interaction": "site is understood here as a generative locus of individual and collective identities, actions, and histories, and the unfolding subjectivity awaits the specific insights generated by the singular coming-together" (2011: 139). For Estelle Zhong Mengual, the term "art in common" (or "*art en commun*" in French) refers to projects created outside of the artist's studio, involving an "exteriorised activity that takes place in the world" (2019: 55) over a period of time. It is a "co-production" (2019: 55) where the work is no longer the "fruit of the artist *alone*" (2019: 11). These are all areas where the conditions and possible forms of collective action can be reinvented, where artists, such as Lone Twin, Jeanne van Heeswijk, and Jeremy Deller, have developed activities (altruistic giving, storytelling, historical re-enactments, bricolage) in places of everyday life.

Kester adds that "this entails a movement between immersion in site and distanciation from it" (2011: 139), a necessary condition to avoid the spectre of instrumentalisation. Indeed, there is always the risk that a work which analyses a site can also be used for institutional functions that impose constraints or be conceptualised as art tourism or as a social impact or economic development project. Kwon warns of the extent to which some *in situ* works have become a mere method or style, received as superficial criticism and reassimilated into the art world as commodities.

However, Kwon also highlights that the notion of the site, initially conceived as fixed and limited by a geographic place or an institution, is gaining greater flexibility and freedom, with the relationship between subject/object and the place becoming less stable:

> The site in now structured (inter)textually rather than spatially, and its model is not a map but an itinerary, a fragmentary sequence of events and actions *through* spaces, that is a nomadic narrative whose path is articulated by the passage of the artist.
>
> *(2002: 29)*

Since the end of the 1960s, *in situ* practices have undergone several transformations and more than one reassessment of a piece of art's autonomy. Recent thinking on the concept of a site has sought to encourage different fields to share the space and a commitment to shared experiences on site, often integrating everyday life through constructing situations. *In situ* art aims to reinvent new forms of commons and community within subjective and collective spaces. Edgar Morin expresses this necessity to continually reconceptualise our milieus and their connections from a transversal approach of complexity:

> Complex thought is a way of thinking that connects. Complex ethics is an ethics of *reliance* [a French term coined by Morin meaning rebinding]. [...] We must, for each and every one of us, for humanity's survival, recognise the need to connect: connect with our own people, connect with others, connect with the Homeland Earth.
>
> *(2004: 248)*

Through the situations they construct, community art projects strive for these connections, manifesting the desire to eschew the art market and make a socio-political commitment based on a relational process of co-production with participants who were once audiences (Kester 2004).

Three Key References

Kester, Grant H. (2011). *The One and the Many. Contemporary Collaborative Art in a Global Context*. Durham and London: Duke University Press.

Poinsot, Jean-Marc (2008). *Quand l'œuvre a lieu. L'art exposé et ses récits autorisés*. Genève: Les Presses du Réel, 80–109.

Zhong Mengual, Estelle (2019). *L'art en commun. Réinventer les formes du collectif en contexte démocratique*. Dijon: Les Presses du Réel.

41

SOCIALLY ENGAGED ART

Isabelle Ginot

The long history of artistic practices that are driven by social engagement crosses all disciplines, perhaps in an unequal manner, and has produced a wide variety of forms. It relates to the ongoing critique of a dualist definition of art and artists as if they were separated from the real world, as well as the debates surrounding reality and representation. Forming part of this history are all practices that are established as a criticism of the artistic institution, those that reject how institutions and the market frame art forms (museums, galleries, theatres, concert halls), and those that refuse the definition of art as distinct from the ordinary world (specialised and technical skills of the painter, sculptor, musician, dancer). Any moment in the history of aesthetics in which artists have thrown into question the primacy of the object (the work) over the process could also be cited. Since the second half of the twentieth century, the labels applied to practices that are at once processual and relational, and which could be qualified as "socially engaged" art, have been multiplying: performative, *in situ*, relational, conversational, participatory, contributory. Whilst these practices are also characterised by their interdisciplinarity, the field of plastic arts is perhaps where the initiative has predominantly been taken.

Pablo Helguera outlines a number of criteria for socially engaged art: practices must be "actual, not symbolic" (2011: 5), understood not as simply representing social questions but acting upon and within them. All socially engaged art projects are about social interaction. In other words, there is no artistic object without interactions and artistic processes. Socially engaged art is transdisciplinary in that it seeks to respond, through art, to a question that is not ordinarily considered as falling within the domain of art. As such, creating a new school, considering the usages of a territory or land, examining social

DOI: 10.4324/9781003455523-42

barriers between two groups of people, and engaging with questions of urbanism can all be the subject of socially engaged art. However, Helguera insists that it is not a case of renouncing art in favour of a purely social engagement, but rather maintaining the unresolved tensions and contradictions that are inherent to this double anchorage in the social and the aesthetic. According to Helguera, this rooting in a social terrain that is not reserved for art and artists also makes socially engaged art necessarily a pedagogical – or transpedagogical – art. It thus becomes a site of learning for the participants and cannot eschew a reflection on pedagogy.

In offering such a restrictive definition of socially engaged art, Helguera sets out the requirements that situate the practice within the vast landscape of contemporary, processual, relational, and participative practices that have flourished from the twentieth century onwards, whilst also avoiding the trap of an aesthetic and political catchall that could result in interdisciplinarity, a lack of specialisation, and the inclusion of themes that are not specifically artistic. Yet, a final criterion seems to determine his definition: to be described as socially engaged, an artistic project must have the intention and provide the means to actually transform something in the social world. He cites examples that allow us to better understand what this means: for example, a collective of artists who decide to create a school that is different from mainstream education can consider their project a work of art. Yet, Helguera would only deem it to be "socially engaged" on the condition that the school would not become solely for children and teachers from the artistic and intellectual community that conceptualised it.

In brief, a socially engaged project must not only respond to a question or an issue from an area outside of art, but it must also, perhaps above all, be established as an instrument to critique one or more social norms embedded within the given situation. This distinction could be used to examine many "ecologically engaged" art projects. Indeed, there are numerous art projects that wish to distance themselves from dominant artistic conventions, such as the centrality of the author and the work as object. Yet, these so-called avant-garde practices rarely show interest in the segregative operations present within the world of art and its intellectual culture. Thus, many processual works, which foreground the link with nature and the intimate experience of sensing, develop apparatuses of access that are only available to a learned elite.

In contrast, other artistic projects are less preoccupied with their position in the art world than their place in the wider world. Within the field of dance, there are many artists who consider their artistic work as a common space to encounter a diversity of corporealities as shaped by social inequalities. Here, I am not referring to extravagant and quasi-commercial stage productions whose success in the institutional dramatic network plays on the surprise of bringing different body types to the stage. For example, the choreographer Jerôme Bel's *Disabled Theater* (2012), created for the Swiss company Theater

Hora for actors with learning disabilities, toured the world's biggest theatres. Rather, I am thinking of the infravisible work of artists who take up residence in neighbourhoods, refugee camps, care homes, and shelters and offer to share – to pool – knowledge about the body and sensations through projects that invariably lie somewhere between artistic work, care work, and political engagement.

Like the numerous proposals for art walks, it is about politicising the sensory, extracting oneself from the dominant notions of the body's naturality and its systems of perception, and understanding that listening to the world would not be a return to a primitive sensoriality located in the body that escapes the influences and traumas of culture and social inequalities. If artists can contribute knowledge of the senses, of listening to, moving within, and perceiving the milieu, these tools cannot avoid coming into contact with other forms of knowledge about the body, acquired in the struggle that is the common fate of those who migrate, those who live outside functional or cognitive norms (so-called "disabled"), and those who are considered as unproductive.

Henceforth, it is necessary to recognise the reciprocity of learning and knowing. We must leave behind the field of art to examine what art can do to dissolve, or at least suspend, the numerous social, economic, architectural, and geographical borders that separate bodies and structure their discrimination.

Three Key References

Bishop, Claire (2012). *Artificial Hell: Participatory Art and the Politics of Spectatorship*. London: Verso.

Helguera, Pablo (2011). *Education for Socially Engaged Art: A Materials and Techniques Handbook*. New York: Jorge Pinto Books.

Marchart, Oliver (2019). *Conflictual Aesthetics: Artistic Activism and the Public Sphere*. Berlin: Sternberg Press.

42

SOUND AND SOUND MILIEUS*

Makis Solomos

What Is a Sound?

Imagine a Sunday in the countryside. It's spring. As you walk, you slowly become aware of the sound of an aeroplane crossing the sky. You don't remember when it started. Through noticing the aeroplane, a soundscape begins to emerge: your attention is drawn to the cawing of crows nearby, a chainsaw in the distance takes on new meaning. Even the wind rustling through the trees is part of this complex patchwork of sound. But it also affects your mood, which, in turn, makes you understand the interweaving of sounds, their rooting in the environment, and your relationship to the land and the sky differently. Sound is thus defined as a network of relations: to other sounds, to the surrounding space, to the subject who listens.

Leaving aside the phenomenological description, the basic, physicalist definition of "sound" tells us that it constitutes a curve of pressure created by a vibrating object and transmitted by molecules from the environment in which it is communicated. Sound also designates the auditive sensation that arises from this vibration. As a vibrating object that is communicated in an environment and becomes an auditive sensation, sound is, evidently, manifested through a fabric of relations.

* This entry revises and retranslates material previously published in a 2018 article entitled "From Sound to Sound Space, Sound Environment, Soundscape, Sound Milieu or Ambiance", in *Soundings and Soundscapes*, Sarah Kay and François Noudelmann (ed) Special Issue of *Paragraph: A Journal of Modern Critical Theory* 41(1): 95–109. We kindly thank Edinburgh University Press for allowing this material to be republished here.

DOI: 10.4324/9781003455523-43

If the sound is manifested through a fabric of relations, one might then wonder: does something that can be called a sound actually exist? The response is: yes, but only when it is captured by and stored on a device. Indeed, it is not sufficiently recognised that a recording constitutes a sort of birth certificate for a sound. It is not that sound did not exist before recording but conceptualising a sound in and of itself is to think of it as a result, something fixed and repeatable. Traditionally, sound has been presented as a sort of epiphenomenon that could be ignored, as the famous ear training, which is a core component of any formal music education, typically does.

At present, there is a move away from the culture of recording that has shaped three or four generations of listeners and musicians. The recorded sound is but one possibility amongst many others. Current developments in technology allow for the borders of what is called "sound" to become more fluid. For example, sound spatialisation technologies make us aware of sound's intrinsic relationship with the surrounding space. Many musicians are therefore working with the idea that sound does not exist outside of a given space, a place.

Sound Milieu

For this reason, the use of compound expressions that connect the word "sound" to a notion of place are becoming increasingly common: sound space, sound environment, soundscape, sound milieu, sound ambiance. These expressions seek to assess precisely how sound is understood, not as an object but defined in relation to a network of relations. Amongst the possible options, sound milieu seems the most promising as it resonates particularly well with the idea of sound as a fabric of relations. The term is starting to be employed in the fields of geography, sound studies, sound design, and ethnomusicology (see Augoyard and Torgue 1995; Roulier 1999; Maeder 2010; Guillebaud 2017; Böhme 2017) where its usage often remains generic as a synonym for sound space, environment, or ambiance.

To understand what constitutes a "sound milieu", an exploration of the concept "milieu" within French thought is required. Dating from the seventeenth century at least, the noun "milieu" has two meanings relevant here: first as a surrounding or "environment", its general English translation, and second as "between", "centre", or "middle" as in "mi-lieu", literally mid-place. The term "milieu" was first used in the physics of Pascal and Descartes before acquiring a meaning in biology (Lamarck, for example) and finding a use in sociology (Compte, Tarde, Durkheim), thereby becoming of interest to geography as well. Throughout the twentieth century, the concept was developed in the philosophies of Merleau-Ponty, Canguilhem, Ellul, and Simondon as well as, more recently, Augustin Berque.[1]

To explore the concept further, let's take a closer look at the German naturalist and biologist Jakob von Uexküll, whose work influenced French philosophical thought, and Simondon. According to Giorgio Agamben:

> Where classical science saw a single world that comprised within it all living species hierarchically ordered from the most elementary forms up to the higher organisms, Uexküll instead supposes an infinite variety of perceptual worlds that, though they are uncommunicating and reciprocally exclusive, are all equally perfect and linked together as if in a gigantic musical score, at the center of which lie familiar and, at the same time, remote little beings.
>
> *(2004: 40)*

To name these multiple worlds that correspond to each animal species, Uexküll, in *A Foray into the Worlds of Animals and Humans* (2010), develops the concept of *"Umwelt"*, literally meaning the surrounding world, translated as "environment" in English and "milieu" in French. Each species lives in its own *Umwelt*: species give meaning to their *Umwelt* and the *Umwelt* imposes its attributes on its species. An *Umwelt* only makes up a small part of the environment (*Umgebung*), comprising only of the objects with which an animal interacts. These interactions are based upon the perception–action loop and produced by the physiological senses, which introduces a species-specific subjectivity to them. Uexküll cites the famous example of the tick that lies in wait for its prey: "the whole rich world surrounding the tick is constricted and transformed into an impoverished structure that, most importantly of all, consists only of three features and three effect marks – the tick's environment" (2010: 51). Whilst Uexküll devotes little attention to human environments, he concludes his book with an example that attempts to support his hypothesis of parallel worlds: "The environments of a researcher of airwaves and of a musicologist show the same opposition. In one, there are only waves, in the other, only tones. Both are equally real" (Von Uexküll 2010: 135).

Reference to Uexküll helps clarify the concept of "sound milieu". The environment mentioned above in the physicalist description of sound is one and the same: the space in which vibrations, provoked by a moving object, are disseminated, giving birth to the sound. Yet, a reading of Uexküll invites us to move away from the physicalist perspective that defines environments as objective spaces in order to consider them as "milieus of". Milieus, whether collectively or alone, do not exist as states of matter (gas, liquid, solid): they only make up one part of the physical environment, which is determined in relation to the subject that interacts with them.

Taking the listener as the subject, their integration into a sound milieu is through the act of listening. The result of this interaction forms what we call a sound. A sound therefore is not an object: it is neither the vibrating object, nor the general space of vibration. Sound is the product of the listener–milieu interaction. Rather than being about contemplating the objects that one encounters, listening thus becomes an immersion within a milieu. Moreover, given that a milieu constitutes but one part of the environment, listening is not a detached form of perception centred on one single sense whilst anaesthetising the others; rather it weaves the relationship between the subject and a specific milieu, the sound milieu which coexists alongside the visual and tactile milieus.

Simondon's concept of milieu draws on the two meanings mentioned above. To give meaning to the extremes, to understand the individual and the environment (rather than the subject and the object or the spirit and the world), it is necessary to begin from the mi-lieu, to understand the relation itself (see Petit 2013). Simondon's definitions can be transposed onto the concept of a sound milieu. By way of an example, we can transform two of his explanatory sentences through substituting the following words: "individuation" with "becoming-sound", "being" with "sound-space", and "individual" with "sound". First, the phrase "we must start with individuation, with the being grasped in its center according to spatiality and becoming, and not with a substantialised *individual* facing a *world* that is foreign to it" (Simondon 2020: 11) becomes:

> We must start with becoming-sound, with the sound-space grasped in its centre according to spatiality and becoming, and not with a substantialised *sound* facing a *world* that is foreign to it.

Making the same substitutions, the second phrase "individual and milieu should be taken only as the extreme, conceptualizable, but not substantializable, terms of the being within which individuation takes place" (Simondon 2020: 367–8) becomes:

> Sound and milieu should be taken only as the extreme, conceptualizable, but not substantializable, terms of the sound-space within which becoming-sound takes place.

Since sound milieus are largely created by technology, Simondon's work is especially valuable on account of his concept of a "technical milieu", the crowning achievement of his argument.

Notes

1 For more on the history of "milieu", see Canguilhem's seminal work ([1952] 2008) that draws an extensive genealogy of the concept, as well as articles by Petit (2013) and Bottiglieri (2014).

Three Key References

Böhme, Gernot (2017). *The Aesthetics of Atmospheres*, Jean-Paul Thibaud (ed) London: Routledge.

Guillebaud, Christine (ed) (2017). *Toward an Anthropology of Ambient Sound*. London: Routledge.

Solomos, Makis (2023). *Exploring the Ecologies of Music and Sound: Environmental, Mental and Social Ecologies in Music, Sound Art and Artivisms*. London: Routledge, chapter 1.

43

SOUND ART

Susana Jiménez Carmona, Carmen Pardo Salgado, and Matthieu Saladin

Many practices associated with sound art relate to a materiality of sound, which draws on a broad definition of sculpture (Schulz 1999: 9; Licht 2007: 11). This expanded usage of the term is semantically linked with what Roberto Barbanti locates as the place of the medium, and in particular to the notion of intermedia developed by Dick Higgins, which "refers to a condition of substantial indeterminacy of the medium as it is: whether it be language, object, actual mediation, or a specific technical tool" (Barbanti 2009: 47–8). The first manifestations of this way of thinking can be observed both amongst the historical avant-gardes and at *Untitled Event*, a "happening" organised by John Cage and Merce Cunningham at Black Mountain College (USA, 1952). This happening presented multiple events and focal points that were not in conflict with each other as no dominant meaning was said to be imposed. *Untitled Event* operated on an indeterminacy, a non-intentionality.

Higgins, who studied with Cage and participated in the Fluxus Festival (Wiesbaden, 1962), first used the term "intermedia" in 1964. He subsequently applied it to creations that fall between recognised artistic media or combine art and non-art. Situated between categories, these practices reflect the social problems that characterise the period and challenge compartmentalised approaches to art (Higgins 1966; Toop 2000: 107).

Moving between media implies an interest in the middle ground that lies outside of established hierarchies and thus supposes:

1 A shift away from the paradigm of vision that has dominated thought and practices in the Western world since ancient Greece.

DOI: 10.4324/9781003455523-44

2 A strong connection to the timespace framework, which often sets aside memory in favour of forgetting. Salomé Voegelin uses the compound "timespace" to refer to this framework. In the production of timespace, place is considered within a socio-political dynamic, which includes viewing the subject as an aesthetic-political being (Voegelin 2010: 265; LaBelle 2010: xix; Pardo 2017).

3 Being available to listen to the milieu from a perspective that is not compartmentalised by the anthropocentric hierarchies that can only prevent relationships with other living beings and those described as "inaudible". Understood in this sense, sound art becomes sound ecology.

Non-anthropocentric

Can humans really listen to non-human sound worlds? Are we really capable of situating ourselves according to other modes of listening, without transforming them into human forms, in other words avoiding anthropomorphism or anthropocentrism? An anthropocentric perspective humanises other living beings by hearing the sounds they make as voices filled with intention, more or less understandable to us and produced for us humans. Indeed, anthropocentricism relies on the notion of listening to other living beings as sound objects, sterilely disconnected (one from another, from a milieu, from a temporality) by so-called objective (transparent and invisible) subjects who believe they are able to escape their limited, fragile, dependent, and mortal human bodies.

Perhaps the issue resides in a desire to situate oneself within other forms of listening, to claim to identify with, understand, and possess knowledge about others. Sound artists have attempted to decentre the *anthropos* without seeking identification or erasing alterity, but by showcasing and amplifying the sounds their cohabitants make. Works like Jana Winderen's *Rats – Secret Soundscapes of the City* (2017), Félix Blume and Sara Lana's *Mutt Dogs* (2017), and Tomás Saraceno's jam sessions offer different and sometimes controversial experiences of listening, where we become living and dying beings amongst other living and dying beings who are very different from ourselves.

We can thus identify a non-anthropocentric approach to sound art, which is based on an attentiveness to other ways of living. These (human) practices necessarily involve the use of microphones, sound treatment, and other technologies that are conceptualised as foreign yet embedded interfaces with the milieu or "*Umwelten*" (Von Uexküll 1957). It is not a case of situating oneself "in" or "in front of", but "with" and "between". A microphone placed in *between*, in the contact zone between subjects and objects, between speaking subjects and subjects that are spoken to, can act as a multifaceted link between forms of listening, relationships, and heterogeneous and mutant alliances. This technique facilitates the amplification of non-human sounds and (self-)composition with others.

The Inaudible

Lending an ear to non-human sounds evidences a broader approach to listening that runs through the history of sound art, as well as underpinning entire swathes of contemporary creation: being attentive to the *inaudible*. Listening thus makes available what at first seems denied. With an ecology of attention that leads to discrete and fugitive worlds, it reconnects us to all that populates them, to what we describe, undoubtedly often too hastily, as silence, and to what the political orders and real-life divisions set aside or attempt to relegate to the margins. Yet, this attention just as often leads to what covers up and masks ambient noises: the media and other forms of authorised broadcast. In fact, so much of what is considered "almost nothing" is actually almost everything. We can cite how Max Neuhaus and his inaudible installations worked on reviving contextual attention; how Alvin Lucier examined the complex acoustic relationships that shape our perceptions in a given milieu; how Christina Kubisch and her critical ear unearthed the electromagnetic fields that form a backdrop to our daily lives; and how Ultra-red investigate the silencing carried out by neoliberal governance.

In all cases, the inaudible describes not only what is situated below or beyond the audible spectrum of the human ear, but also what appears to escape the hearing of a given group. The mute noise informs us, by means of the negative, what our societies consider as pertaining to the audible, its representations, and its cultural and ideological inscriptions. By paying attention to the inaudible, sound practices contribute to a redefining and displacing of the audible's boundaries, bringing to light once again their contextual, circumstantial, and cultural nature. If the audible is a spectrum with blurred borders, it is this porosity that feeds the "spectral nature" of its opposite. By amplifying the sounds of spaces or the technological mechanics that regulate our daily lives, by transposing the sounds of insects, plants, minerals, or objects situated below or beyond the human ear's perceptive threshold, or even by listening to the voices politically considered as "silent", sound artists make us (re)listen to the background noise against which what we readily accept as audible stands out. These practices can be described as "political" in the sense that they reconfigure the perceptive thresholds and the social divisions they suppose.

Three Key References

Jimenez Carmona, Susana, and Carmen Pardo Salgado (eds) (2021). "Aperturas y Derivas del Arte Sonoro". *Laocoonte. Revista de Estética y Teoría de las Artes* 8: 49–56. URL: https://ojs.uv.es/index.php/LAOCOONTE/issue/viewIssue/1486/313

Saladin, Matthieu (2020). "The Inaudible as an Effect: Tactics of Sound Erasure in Max Neuhaus", in *Sound Unheard*, Anne Zeitz (ed) Special Issue of *Kunsttexte.de* 1.

Solomos, Makis, Roberto Barbanti, Guillaume Loizillon, Kostas Paparrigopoulos, and Carmen Pardo (eds) (2016). *Musique et écologies du son. Propositions théoriques pour une écoute du monde*. Paris: L'Harmattan.

44

TECHNOLOGY AND AN ECONOMY OF MEANS

Agostino Di Scipio

In a world where digital media networks and highly complex technical infrastructure forge processes of subjectivation and the collective of human relationships, what can art do? What position and which direction can art take when faced with the labyrinth of technological mediations in which we live? What about, for example, music or sound art in general where all production and perception of sound take place in and through multiple interlocking layers: electroacoustic, digital audio, and telematics? This entry attempts to address these questions with reference to the current phenomenon of a widespread and profound impoverishing of sensoriality.

As well as producing beauty or pleasure, art also invents and transforms the link between historically determined ends and means. Over the course of the twentieth century, artists learnt to make use of new technological resources through several conceptual and operational paradigms (avant-garde, experimentation, research, cultural production). They have acknowledged a profound disruption: technological developments were no longer limited to defining new forms of instrumentality, the scope of and use of technical means, as they came to radically reconfigure our common living environments as several philosophers, from Heidegger to Simondon, from McLuhan to Guattari, have noted and discussed. Technical networks have become conditions of possibility for psychic and social processes, tending to redraw the scope of human relationships with physical reality and with what is considered "nature": we live in the age of technology's "becoming-milieu".

As Bernard Stiegler (1996), following on from André Leroi-Gourhan, highlights, a fundamental link in the anthropo-eco-logical order unites humans, as well as other living beings, with their specific milieu and the general environment. The

DOI: 10.4324/9781003455523-45

environment, itself a constructed space, designates a spatial niche where the human subject exteriorises and distributes their cognitive faculties and capacity to act using different machines and devices. This, according to Erich Hörl (2013), is the root of the current technological condition, where human beings dwell in environments almost entirely composed of interconnected electronic networks. This condition raises the important ecological (and political) question that concerns us all: how to construct one's milieu without destroying that of others?

The prevailing economism considers ecology in terms of technological development and planning. Yet, technologies urgently need to be viewed from a relational and ecological perspective. The paradigm must be reversed: instead of approaching the environment and ecology in economic and technocratic terms, the economy and technology need to be considered through a relational and ecological consciousness. By adopting this paradigm, creative artistic practices may act as forms of praxis capable of preserving and articulating humans' ability to share social and sensorial experiences.

Whilst economic growth and technological opulence have forged the dogma of "indefinite" progress without definition, end, or goal, artists can act in other ways: they can appropriate their means of action, their workplaces, and reshape them according to an economy of means that is at once moderate and appropriate. By renouncing the creative demiurge's *beau geste* and opening up to an ecological way of thinking, creative practices can escape economism and construct a praxis in which communication no longer means transmitting and selling information but acting in common and creating shared conditions of action, knowledge, and sensorial perception.

More desirable than an indefinite development of technological competencies is the capacity to discern a necessary balance, a harmony between means and ends, a harmony always relative to a specific framework of action, to a given situation, that is in relation to changing but real material contexts. The most technically advanced feature of creative work consists of individuating such a balance: goals do not impose any predetermined means (the means are to be sought out, appropriated, or designed from scratch), just as means do not impose any predetermined goals (the goals are always to be redefined and negotiated). Ends and means are reciprocally, dialectically intertwined: ends and means each emerge from one another.

This is the only harmony that counts from now on. It is not about long gone musical tradition (Pythagorean, tonal), nor about aesthetics in general. It is a matter of rethinking creative praxis as a search for and a materialisation of "bonds of liberty", which are to be reconsidered and experienced according to different media and contexts. Experiencing bonds of liberty means providing sensorial evidence of the dialectic between autonomy and heteronomy, perceiving the room for manoeuvre or lack thereof between acting and being acted upon, as well as the possible (even inevitable) clash between these two. In contrast to the dominant aesthetic-technological discourses, artistic practices would develop their own technical conditions of possibility, thus bearing witness to the fact that, fortunately, *not* everything is possible.

The aestheticisation of technology and the technicalisation of aesthetics, so characteristic of the dictatorships of yesteryear and the current prevailing hegemonic economism, constitutes the apparatus by which disinterested appreciation (in the Kantian sense) of artworks and the ideology of indefinite progress go hand in hand despite appearances. Confronted with this, artistic practices would seek rather to reinvent the conditions of the sensorial and to start a battle for a consciousness of the sensorial (to appropriate its own perceptions) and for sensodiversity (akin to biodiversity). Such an aesthetic war requires recentring oneself on the conditions of production or, better still, on the ecology of media and creative processes as means of individual and collective subjectivation. There is no freedom of expression without freedom of action.

Conceptualised in this way and regardless of specific media and artistic languages, the distinctive trait of creative practices is an opening up and expanding of the (very reduced) room for manoeuvre between action and perception. This attempt is situated at the very heart of a transition based on a relational and hybrid direction: a techno-eco-logical perspective, which leads us to feel the connectivity rate or the ecosystemic density in the circular relationship presented here:

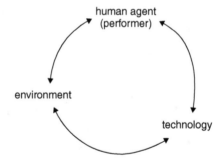

The double-headed arrow means "acting on, doing something to" *and* "receiving, submitted to something".

An aesthetic praxis in transition makes perceptible the non-dissociability of elements in this interaction loop. Yet, it does not rely solely on the dynamics of that threefold interactional field, but also on how the modes of interaction themselves depend upon each other:

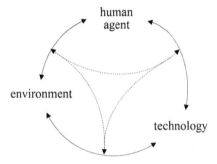

In one respect, sound constitutes an exemplary medium for understanding and living that loop of relations as long as it can be worked on, heard, and experienced as the audible exchange between human agents (gesture, body, ear) and the various non-human agents that structure the surrounding physical space. This constructed space is thus an acoustic space as well as a place loaded with social and cultural connotations. The sonorous medium constitutes a hybrid ecosystem, a fabric of different informational and energetic flows (mechanical, electroacoustic, digital audio), forming a mixed assemblage of the human and non-human (Simondon 2012). Sound makes audible the interdependence of multiple agents, their co-dependence and mutual determination. It accords a phenomenal reality to the systemic notion of "dependant autonomy" as discussed by Edgar Morin (1977).

No longer "audio", nor "sound objects", sounds thus appear to our listening capacities as relational events, as phenomena emerging from and belonging to a network of material, situated interactions. As a milieu of auditive and corporal perception, sound communicates – or disseminates – the responsibility and the conditions of possibility in which sonorous events emerge: a responsibility one can take or leave, appropriate or delegate, capture or miss; a perceptible responsibility, felt in the body and aurally understandable. Sound takes place in a sensorial register that is intrinsically political. Even if it is experienced as background noise (so-called silence), sound always demands an active and participative perception, an attitude of "implicated listening" (as opposed to "reduced listening"). Being physically situated *in* (shared environments, shared circumstances), *between* (sonorous sources), and *amongst* (other listening subjects), whoever is listening is, by definition, interested in or indeed implicated in what is listened to.

Whatever the phenomenology of the sensorial experience in question, it is a case of linking, coupling, weaving together, and even contrasting the components of a larger (eco)system by making perceptible their dynamic relationships or lack thereof. Art practices show that mediality and sensoriality are intimately connected and create a feedback loop through which the interaction between humans and their milieu must be worked on (Parikka 2019).

Today, as milieus are increasingly technologised, humans have an increasing tendency to turn themselves into cyborgs. It is necessary then to experiment in other ways, to de-structure and reinvent a techno-eco-systemic phenomenology to live with. It is necessary to act on and within the technological condition (sometimes also called the post-human condition) and the thousands of ecologies that proliferate there. The labour of art thus turns into a praxis that is biopolitical in character: it develops a strict economy of means and opens up to a sensory ecology that reveals the relational (biological and cultural) conditions of one's belonging to a shared living environment.

Three Key References

Di Scipio, Agostino (2021). *Circuiti del tempo. Un percorso storico-critico nella creatività elettroacustica e informatico-musicale.* Lucca: LIM.

Solomos, Makis, Roberto Barbanti, Guillaume Loizillon, Kostas Paparrigopoulos, and Carmen Pardo (eds) (2016). *Musique et écologies du son. Propositions théoriques pour une écoute du monde.* Paris: L'Harmattan.

van Eck, Cathy (2016). *Between Air and Electricity: Microphones and Loudspeakers as Musical Instruments.* London: Bloomsbury.

45

TERRITORY

Ludovic Duhem

We must return to the concept of territory. With the catastrophic state of the world, it seems unavoidable. To evaluate such an imperative, we must first define what is meant by "territory". Five conceptual fields can be identified: territory can be understood as a geographic area; a political border; an economic network; a point of origin; and a living milieu. These five fields are not mutually exclusive. They can overlap or enter into conflict with each other, fuse together, divide each other up or interlink, and appear and disappear over the course of time.

As a geographic area, the territory designates a part of the Earth's surface. More precisely, the territory is an area of land whose limits are largely identified, marked, and constructed, thus forming a recognisable unit that is distinct from other areas and the rest of the known world. If a territory's boundaries can be incomplete or change over time, a minimum of elements must exist within a tangible space, or an outline corresponding to the represented space is required, so that it can feasibly appear as a territory. These elements can be morphological (e.g., reliefs), artificial (e.g., ramparts), or symbolic (e.g., religious signs); they often coexist without always overlapping each other. Before becoming a space of knowledge, the geographic territory is stable, objective, flat; one does not experience it.

As a political border, the territory designates an area occupied by a group of humans who submit to their own formalised authority. This occupation is ratified according to a spatial organisation demarcated by the physical borders that confer its unity and its identity, and by a claim to sovereignty over the space that is occupied and demarcated. Sovereignty is exercised over all the individuals who live there according to the same rights and obligations.

DOI: 10.4324/9781003455523-46

The territory is transformed through conquests and invasions, treaties and conventions, revolutions and civil wars, partitions and reunifications. Unity and coherence must be assured between the physical space of settlement, the legal space of exercising law and sovereignty, and the symbolic space of representing a people and the authority that derives its existence from the territory. Therefore, the reality of political territory is not only managing infrastructure and representing an authority, but governing individuals by means of the space as well.

As an economic network, the territory designates a collection of hubs and links connecting a given space through the flow of materials, goods, and capital. These hubs are commercial centres where economic power is concentrated, where exchanges converge, or where value is fixed. Hubs can take the form of a marketplace, a stock exchange, a trading post in a foreign country, or an entire city designated as an economic capital. Links between hubs develop through large- and small-scale infrastructure (from a simple route to a digital spatial network) and have a direct influence on the representation and reality of the territory. Above all, the economic territory is an abstract and planned space, a nondescript system available for circulation, appropriation, exploitation, and representation.

As a point of origin, the territory designates the starting point of a movement, the birthplace of an individual or a people, or the identifiable source of a symbolic quest. At best, it indicates what had to be left in order to come back better, like a home whose influence can only be felt after a departure, a return that only acquires meaning after an arduous journey, or succour that can only be offered in the long term after perilous wandering. At worst, it calls for a retreat into the authenticity of origins, the veracity of identity, and the inalienable property that only the native land of a people who share ancestors and the blood of those sacrificed can express and preserve. There is a substantial risk of being seized by nostalgia for what has been definitively lost or the regressive illusion of returning to the beginning. Worse still, succumbing to a fixing of identity opposes any alteration to property, any estrangement from the sovereign territory, and any deviation from the law of the land. In contrast to this inward retreat, a territory that opens up is one without origins, which welcomes the uncompromising and dynamic diversity of worlds.

As living milieus, the territory designates a complex and multi-scale collection of interdependent systems whose active relationships define the conditions of existence and development of life there. Living milieus are not objective parts of the natural world, nor natural environments detachable from living organisms. To understand them requires a decentring that shifts the perspective to these relationships, rather than an individual placed at the centre or the exterior. A living milieu is formed from relationships, not logical links between concrete terms, but precisely from what is concrete, active, and constitutes the terms (Simondon 2015). As such, no living individual has their own milieu:

they exist according to the milieu just as the milieu exists according to them. The territory is a network of living milieus where individuations spread from one to another, maintaining the potential and diversity of life through vital activity. This network of living milieus is replicated in a superior network of places, which express the key characteristics of living milieus. A territory that has become a world thus represents the true and common meaning of the relationship to a specific place.

Today, this relationship to place has deteriorated, diminished, or even been completely severed due to lifestyles founded on the utilitarian, mercantile, and ostentatious concept of territory that industrial modernity and global capitalism impose. This rupture in the relationship to place is a decisive condition in the territory's development: as it becomes increasingly concrete yet imperceptible for the inhabitants, the land becomes increasingly available for all forms of exploitation, construction, and destruction. The territory thus is no longer a living space to defend, express, share, and care for but is the object of a competition for exclusive property. For the territory to be anything other than an abstraction and an alienation, it needs to exist as a real, living space, an arrangement of unique places that cannot be appropriated or measured, through which lines of deterritorialisation and reterritorialisation run and where the human and non-human worlds are brought back together by transforming each other.

This role can be taken up by an art of the territories, namely an art that is capable of re-establishing the relationship to place, of making it perceptible, of expressing it through unique experiences, and of taking care of it through situated practices. This is necessarily an art of criticism (of its possible conditions and of the established order), an art of situations (of the system of eco-techno-symbolic reality), and an art of living milieus that cares for and celebrates a diversity of sensibilities, imaginaries, and cultures. The art of the territories is not an economic sector, an established discipline, or a market of cultural goods. It is a living art and an art of living spaces, inspired by Bookchin's social ecology, Berg's bioregionalism, Guattari's ecosophy, and Berque's mesology. Therefore, it cannot exist, nor continue to create and bring together worlds, without questioning its own responsibility in the ecological disaster, nor can it refuse to fight on the ground against the aesthetic, social, and ecological poverty that removes all the joy of being a living being amongst living beings.

Three Key References

Duhem, Ludovic, and Richard Pereira de Moura (2020). *Design des territoires. L'enseignement de la biorégion.* Paris: Eterotopia.

Paquot, Thierry, and Chris Younès (2009). *Le territoire des philosophes. Lieu et espace dans la pensée du XXe siècle.* Paris: La Découverte.

Simondon, Gilbert (2015). *L'individuation à la lumière des notions de forme et d'information.* Grenoble: Million.

46

THEATRE

Eliane Beaufils and Julie Sermon

As this entry will interrogate how "theatre" can participate in the artistic dynamics of transition, we begin by noting that, since the 1990s, a rising number of stage artists no longer fall under this signifier. In both practice and theory, forms and hybrid descriptors (such as post-dramatic theatre, devised theatre, performative theatre, performance installation, immersive theatre) have flourished. These alternative ways of doing and describing can themselves be considered as symptoms of the ecological turn, which, deliberately or more intuitively, marks the contemporary arts field.

We are conscious of the transformations that are modifying or redefining theatre's disciplinary contours, and that often render established boundaries obsolete. We nevertheless adopt the position of considering "theatre" as a term and a medium that continues to possess its own characteristics, thereby making it distinct from other artistic practices.

This entry is structured around three areas: stage, action, narrative. We have chosen these three components of theatre for two reasons. First, they may initially seem incompatible with the ecological paradigm and the anthropic decentring that it supposes. Second, they link the mental, social, and environmental dimensions that form the basis of ecosophy's "ethico-aesthetic" triad (Guattari 1989: 31).

Stage

If we endeavour to give the "stage" a general and functional definition, it would be an area cut from the world's space–time continuum. From the ornamental background walls that were progressively built up in Greek theatres, which were originally open to nature, to the contemporary black boxes, this area has

DOI: 10.4324/9781003455523-47

become ever more isolated from its environment. This is theatre's first ecological failure.

The second failure is a dramaturgical and aesthetic, rather than architectural, issue and relates to the entirely peripheral role that space on stage generally assumes. Without being entirely irrelevant, the place of action only has a contextual value: it serves as a framing device, a container (whether abstract or figurative, realist or symbolic) for the action and dialogue that happen on stage, the only elements to take an active part in the performance.

Is theatre structurally unable to give "the non-human environment" agency? Is it possible to understand the stage not simply as a "framing device", a "constant", or a "given", but as a "process" and a "presence" that suggests "human history is implicated in natural history" (Buell 1995: 7–8)? This is not the case. From the 1920s to the 1930s onwards, theatrical avant-gardes have worked to go beyond the "dualism between man, the dynamic element, and the environment (the static element)" by making the scenic space "poly-dimensional" and "poly-expressive" (Prampolini 1926: 103–4); to understand the stage as a "tangible, physical place" that "ought to be allowed to speak its own concrete language", which is "aimed at the senses and independent of speech" (Artaud 2000: 98); and to replace the teleological model of performance with "landscape" plays (Stein 1988: 122) that undo hierarchies between humans and non-humans, decentre the audience's attention, and privilege atmospheric qualities over driving the plot (see Marranca 1996; Lehmann 2002; Fuchs and Chaudhuri 2002). All these artistic intentions and explorations aim to bring the stage to life with all its material and sensorial power.

Yet, making do with the ecological potential of theatre's perceptual, aesthetic modalities is not at all sufficient or satisfactory: "stage craft" hinges on "reconfiguring the coordinates of a field of experience", on the examination of "paradigms of action. What is acting? How to act? How does a cause produce an effect? Who can perform which type of action?" (Rancière 2018: 31, 48).

Action and Actors

As these questions suggest that human agency in theatre has been conceptualised, from Aristotle to Brecht, as closely related to plot. Plot is understood as a chain of actions, causalities, and intersubjective relationships, whether these be fictional links between the protagonists, or relationships between the spectators and with the actors. Located within the city-state and using plot as a form of mediation, theatre is the ultimate place of the *anthropoï*, inviting us to reflect on the place and role that humans occupy.

Because of this, the dramatic genre has become a favoured medium for artists who, at this time of systematic ecological crises, want to ask questions about human actions, about their limitations and their impact. Motivated by a pedagogical or activist desire to warn or raise awareness, to criticise or condemn the excessiveness and irresponsibility of human actions (D. Delorme 2019; Sermon 2021), it is common for artists to draw on subjects that have a didactic, even heroic dimension: from theatre for young people to dystopian civilisations. Other artists, however, prefer to invite the audience to assume their capacity for action through forms that renounce the idea that the actors embody the action and draw on notions of the spectator as "*viveurs*" (Debord 1989: 10), people who live in a conscious manner, or co-creators. Participatory theatre favours a notion of acting as *prattein*, the realisation of protocols, rather than framing and orientating the action around the necessities of the plot (van Eikels 2018: 37). It also calls for the invention of narratives (*Make It Work*, Frédérique Aït-Touati, 2015; *Cracks*, Charlotta Ruth, 2018) or even of ecological and social action in which the spectators have more direct involvement (*1968*, by Proxy, 2018; *Der futurologische Kongress*, Stadttheater Ingolstadt, 2018–19). The indeterminacy of action is a chance to reimagine it. Theatre thus becomes a fundamentally open situation without *arche* (Hetzel 2013: 493), an origin or source of action. This type of theatre, without heroes but not without actors, is a potential ecosophic laboratory that could prepare for future climatic and political agencies (Beaufils 2019b: 145).

Narrative

The question, nevertheless, remains of whether theatre can represent stories and meanings that do not depend on the category of the "subject" but could extend to other forms and processes of subjectivation. Very often, when nature, animals, the elements, and materials have been represented on stage, theatre has tended to anthropomorphise them. Puppetry, on the other hand, has long brought to life objects, non-human presences, to weave a shared history with humans. Now, the contemporary stage is becoming a place of exploration that radically blurs perceptual and ontological categories: objects and materials become, properly speaking, agents that develop spaces, relations, and temporalities according to their modes of being and their own actions (Merabet et al. 2019; Beaufils 2022).

These agents, who are no longer submissive to humans and their agendas, combine multiple languages and potentialities (Benjamin 2000; Latour 2004). They can still convey semes or narratives but also break with constitutive frameworks of intersubjective relationships (dialogue, recognition, performance). Plants, music, or environments thus represent "excessive objects" (Eiermann 2009: 206) or the object as question (Didi-Huberman 1992: 76), which is impossible to grasp entirely. Beyond language categories, these elements also go beyond any narrative; instead of interpersonal or symbolic relationships with the spectators, other relationships could emerge and open up avenues to ecological thought (Morton 2010).

Reciprocally, human agents explore decentrings and deterritorialisations by partially becoming the *things* that escape the subject–object dualism (Eiermann 2009: 264). These processes of impersonalisation are carried out by, for example, abandoning *logos* (reasoned discourse) or depriving movement or speech of address and functionality. Through the absence of *dran*, action as the product of a decision-making will (Guénoun 2005: 81), the undramatised space of the stage once again becomes a situation, occupied by beings who are no longer *anthropoï*. This situation is shared by the spectators and leads to loops of autopoietic feedback (Fischer-Lichte 2008). The spectator is thus also able to experiment and reflect on new relational modes beyond the anthropos (Beaufils 2019a, 2023).

Even when theatre does not abandon dialogue or frontal staging, where spectators face objects, introducing alternative thinking and new narratives that are anchored in a shared situation still has the potential to open up a perspective that serves as a prelude to other beings-in-common with humans and non-humans.

Three Key References

Finburgh, Clare, and Carl Lavery (eds) (2015). *Rethinking the Theatre of the Absurd: Ecology, Environment and the Greening of the Modern Stage*. London: Bloomsbury.

Sermon, Julie (2021). *Morts ou vifs. Contribution à une écologie pratique, théorique et sensible des arts vivants*. Paris: B42.

Woynarski, Lisa (2020). *Ecodramaturgies: Theatre, Performance and Climate Change*. New York: Palgrave Macmillan.

47

TRANSITORY URBANISM

Fabrice Rochelandet

Artists' Need for Places and Spaces

Artists need dedicated spaces to create (workshops, rehearsal spaces, filming locations), promote their creative outputs, and connect with their audiences (concerts, screenings, performances, exhibitions, events, readings) (Meusburger et al. 2009). For many of them, traditional cultural policies and provisions have nothing to offer. Whilst artists must contend with (sometimes very) high property costs in city centres, there are plenty of unoccupied and thus available spaces that could provide opportunities for individual and collective activities. Historically, the legal or illegal occupation of abandoned or otherwise available spaces has offered one solution. The rise of this phenomenon coincided with, on the one hand, the deindustrialisation and economic crises following the 1970s and, on the other hand, how property speculation has accentuated urban transformation and, more recently, led to an increase in city centre property prices, particularly in large European centres, like London or Paris. These two factors have progressively pushed artists to the peripheries. Examples of spaces that artists and collectives have occupied include empty business premises in the UK and Portugal, office blocks in Rotterdam, vacant hotels in Spain, unoccupied residential developments in Riga, disused cinemas in Rome, and industrial warehouses in Paris and its suburbs.

Illegal occupation and the organisation of squats have long established a practice of temporarily using abandoned spaces and properties, whether private or public, that artists occupy without authorisation or paying rent. This offers them the possibility to set up the space in accordance with their needs and to create communities of peers there. At the same time, legal routes to occupy available spaces on a temporary basis are also being developed through

DOI: 10.4324/9781003455523-48

established frameworks between owners and occupants. These occupancy processes are currently being integrated into urban policy with the aim of promoting local development and the regeneration of specific geographic zones. What effects do these legalisation processes and the centralised organisation of occupancies have on the creative dynamics in the local area (see Jacobs 1970; Landry 2008; Florida 2005; Cohendet et al. 2010)? In the following discussion, I will suggest that it risks institutionalising artists' temporary occupancy of spaces.

Regulating Supply and Using Unoccupied Spaces

A large supply of abandoned places, industrial wastelands, office blocks, and former public service buildings is now available for different types of usages: the social and solidarity economy (emergency housing, allotments, workshops), redevelopment for long-term occupancy, or artistic or related activities. How can we analyse the effects of transitory urbanism within this context as a practice and political instrument for the use of abandoned places?

There are four main solutions to regularise the use of unoccupied spaces and, in doing so, address the surplus of such places (Madanipour 2018):

1 Adjusting prices: the implicit idea is that the market auto-regulates itself, so that if there is a surplus of supply, prices will go down until they meet demand;
2 Reducing the supply through local government intervention (demolition);
3 Functional conversion (redevelopment): office blocks or industrial wastelands can be transformed into cultural spaces or residential housing;
4 Temporary or transitory occupancy.

The fourth option becomes essential when the first three solutions are shown to be insufficient, specifically in terms of urban planning and regeneration. It aims to assign one or more usages to unproductive and unoccupied spaces by giving access to actors who do not have the necessary means to carry out a project judged to be useful or profitable and can make use of them. This can be a means of social inclusion that confers a significant role in the process of urban transformation to the residents, local communities, and actors of the associative fabric. The fourth solution therefore includes different institutional arrangements based on specific property regimes (e.g., right to use), including squatting.

To encourage temporary occupancy, different legal mechanisms exist (Patti and Polyak 2015). For example, making the offer of spaces visible and transparent through public databases that allow users to identify abandoned and available spaces and for those interested to contact the owners to negotiate an occupancy.

Temporary Occupancy: Advantages and Disadvantages

Whilst transitory urbanism and the resulting forms of temporary occupancy present numerous advantages, how it is managed has been targeted by critics. On the one hand, it can generate benefits for all parties involved. For owners, negotiating an occupancy allows them to save on guardianship and security costs, whilst also preserving the unoccupied space's value. The temporary occupants access low-cost areas and have the freedom to transform an abandoned place (sometimes ultimately destined for destruction) according to their needs. The local authorities, communities, and the neighbourhood benefit from positive externalities: for example, eliminating the possibility of illegal squatting which can result in delinquency, regeneration of the area, and even improvements to the neighbourhood's image and the conditions of everyday life there.

On the other hand, the negative consequences can outweigh these advantages (Tonkiss 2013; Ferreri 2015). Increasing property prices in the neighbourhoods where these places are located, and the beginning or acceleration of gentrification, can lead to the original population moving away. In addition, the owners of occupied spaces can appropriate (internalise) the social value created by a transitory use of the space without any compensation for the artists, artisans, and event organisers who brought about the improvement to living conditions in the area. To a certain extent, the presence of activists and culturally and socially engaged actors, and the urban regeneration that they engender, can become a target for instrumentalisation in the hands of the most privileged groups. In turn, the social value created by the temporary occupancy of spaces (social and cultural activities, inclusion, projects in working-class areas) can disappear when it comes to an end. Moreover, these processes can be a pretext for reductions in public aid in favour of collectives and cultural initiatives: no additional resources to combat the precarity of artists or homelessness are allocated if local authorities judge that the legal authorisation to occupy a place constitutes sufficient support. In the long term, this institutional process of urban development, which can be described as neoliberal, may affect the local creative and social dynamics.

A Risk of Homogenising the Usages of Spaces

Transitory urbanism can produce negative effects for its occupants through the commodification or privatisation of vacant spaces that were initially common resources, had no competing users, and could be used by artists, militants, and activists without any formal constraints. Far from being an instrument for social inclusion, regeneration resulting from temporary urbanism can produce the opposite effect: the progressive exclusion of inhabitants and poorer artists through the well-known process of gentrification. The development of alternative places for a young bohemian, hipster audience who engage in alternative practices will ultimately attract a small, more wealthy population wishing to live in a trendy place, which makes them more expensive and progressively excludes the original residents and their different cultures.

From my perspective, there is another latent risk whose effects are more difficult to evaluate: the normalisation of artistic and social action to the detriment of the dynamism and diversity of creative forms and social innovation. Initially, occupancies were the initiative of the artists and activists who identified, occupied, negotiated, and renovated unoccupied spaces by transforming them into artist-run spaces, third spaces, or alternative places. If a policy of inviting applications becomes widespread as a way of regulating the use of unoccupied spaces, it could lead to the preselection of projects based on the political interests of local and economic groups of urban operators, rather than those of the occupants. More fundamentally, introducing a process of institutional isomorphism (DiMaggio and Powell 1983), which requires collectives, associations, and intermediaries to apply, may lead to applicants playing it safe, resembling one another, and responding in a very similar way (wording, outcomes, means requested) in the hope of pleasing the decision makers. If applicants are part of the same networks in the art world and the solidarity economy with very strong connections, this is all the more likely. Auto-selection and similar projects that respond to the expectations of urban operators could progressively homogenise the spaces, how they are set up, and what functionality they propose to occupants, as well as having a significant impact on artists' activities (aid, training, creation, promotion) and potentially on their capacity to explore and put into practice new ideas and actions (artistic creations, social innovations). Subverting usages, a diversity of engagements undertaken, and alternative forms of occupation may balance out this process. However, in its transformation of the middle ground made up of alternative cultural places, transitory urbanism in itself would not be any less a tool for the latent standardisation of representations and artistic practices to the detriment of dynamic alternatives that nourish the creative territory.

Three Key References

Ferreri, Mara (2015). "The seductions of temporary urbanism". *Ephemera* 15(1): 181–91.

Madanipour, Ali (2018). "Temporary use of space: Urban processes between flexibility, opportunity and precarity". *Urban Studies* 55(5): 1093–110.

Patti, Daniela, and Levente Polyak (2015). "From practice to policy: Frameworks for temporary use". *Urban Research & Practice* 8(1): 122–34.

48

VISUAL

Claire Fagnart

We can identify several regimes of vision. Retinal vision defines sight on a human scale. With the technical regime of vision, an instrumentalisation of ways of seeing emerges. The technological regime establishes an unprecedented relationship to the visual that is of the same substance as today's hypercapitalism, thereby posing the question of excess.

To observe from retinal distance is to view objects by moving away from them, by separating ourselves from them. Since we can only see what is neither too close nor too far, the term "retinal distance", and therefore the visual observation it offers, applies to a specific distance and not to distance in general. This distance, which allows our organ of vision to see, is adapted to the human eye. Retinal vision implies what Ivan Illich describes as a "natural link between the eye and the world": "naturally", the eye cannot see what is "out of view". For Illich, this mutual relationship between the eye and the world corresponds to a natural measurement that we can understand in terms of proportion. This proportional view defines what is seen on a human scale. The proportion or scale is in line with and relative to each species as it is determined by the organ of vision and the specific needs of each species.

With retinal vision, it is the eye, that is the body, that defines what is considered as the first scale or the human scale in terms of measurement, what Ivan Illich calls the "natural scale".

Retinal vision is based on the combination of the specific distance that allows us to see and the transitive nature of visual contact. Since it is our body, without the support or mediation of technical instruments, that sees, retinal vision is always in the presence of what it is seeing. It then follows that, as retinal vision is attached to the body, it is a synaesthetic form of vision: I see a

DOI: 10.4324/9781003455523-49

landscape and I hear birds chirping. This visual synaesthesia is lost through the use of technical instruments. At the same time, the regime of retinal vision is ephemeral; it is always a lived moment, even if it can be fixed in images on the grounds of representation.

Retinal vision supposes the existence of an observer. The individual becomes an observer when they are situated at the right distance to allow them to see or clearly make out what they are looking at, to examine and visually study it. The retinal observer is an individual in a "'real,' optically perceived world" (Crary 1990: 2), not aided by instruments that modify the distance between them and what is being viewed (e.g., binoculars which reduce distance). A retinal observer emerges at a certain distance from the world, an individual standing at a view-point. Any images resulting from retinal vision (e.g., a sketch) reflect the observer's point of view.

The technical regime of vision is created by using instruments placed between the observer and that which is being observed. These tools, such as binoculars, a spy glass, or a telescope, offer lenses and devices that increase the power of our eyes. Their development in the nineteenth century transformed the concept of an observer. The term "observation" suggests, or even signifies, "vision through use of an instrument" (Illich 2004b: 309). Jonathan Crary employs the term exclusively in this sense.

Technical instruments multiply the distances that form part of our visual understanding of the world: the microscope gives us access to the infinitesimal (our eyes can inspect a cell and see its nucleus); through binoculars or a telescope, we see far-off objects in an accessible way. These devices thus allow us to view what we cannot see. The relationship to the object being viewed, made possible and mediated by a technical instrument, becomes intransitive.

This regime of vision includes technical devices with the same characteristics that reproduce reality in images. The analogue camera generates images that visualise moments we cannot see with our own eyes (e.g., Étienne-Jules Marey and Henri Cartier Bresson). Transforming scenes captured by a camera into images preserves them.

The technical regime of vision, therefore, has the power to expand our understanding of space and time beyond our own sensorial perception. It offers the possibility of escaping the first scale of retinal vision in favour of other scales of vision, representing an enlargement of the spectrum of what we can see or a rupture with the "natural" human scale of measurement. At the same time, the dependence on technical instruments and accompanying detachment from the body leads to a dehumanisation of vision that diminishes its aesthetic character and removes its synaesthetic character. This dehumanisation on sensorial and emotional levels is counterbalanced by an increase in knowledge. The technical regime of vision is very useful to scientific research and has facilitated many advancements in human knowledge.

The technological instrumentalisation of vision introduces the notion of excess. Images are no longer created from what is visible, but from abstractions. These images are constructions or "radical [...] reconstruction[s]" (Crary 1990: 9) of what is visible. Usually the product of millions of bytes of information, they can be constructed from other images, from computer programs that deduce and calculate probability. These images are the result of a complex form of capturing (or pseudo-controlling) the visual. In the passage from technical to technological devices, the relationship to the body is completely erased as there is no longer a sensorial link, even a mediated or indirect one, with the object being viewed. Despite becoming intransitive, this link is nevertheless maintained in the technical regime of vision. It then follows that a technological form of vision is only materialised in images that reconstruct the visible. The retina as a locus of experience, a place of the body and the emotions, is replaced. In its place are images whose autonomy (or detachment from the visible and perceptible) is so absolute that they represent nothing but a belief in what they show.

With this type of image, we are confronted with a visual excess due to the use of extreme scales that can even bring space and time together. Distance is no longer only spatial (we see stars in the sky); it becomes temporal as well (we see the cosmos during the Big Bang). Thus, the ability to produce images from the past, images of objects that existed and no longer exist, was developed. Today, artificial intelligence has the capacity to go even further by producing images of the future, of what does not yet exist.

It is no longer a case, as with technical instruments, of seeing something that one does not actually see but nevertheless exists. Now, we see images of something that no longer exists, does not yet exist, or simply will never exist. The overlap between our human experience and what we see thus disappears. There is an incommensurability in these images, similar to their excessive scale.

This new regime of images cannot be divorced from the precise context in which it emerges. This context is marked by the will to increase control over individuals (socio-political purpose), increase objective knowledge (scientific purpose), and increase profit (economic purpose). The technological regime of vision is in accordance with a hypercapitalist and totalitarian society.

Three Key References

Crary, Jonathan (1990). *Techniques of the Observer: On Vision and Modernity in the Nineteenth Century*. Cambridge MA: MIT Press.

Illich, Ivan (2004). "Passé scopique et éthique du regard. Plaidoyer pour l'étude historique de la perception oculaire", in *La Perte des sens*. Paris: Fayard, 287–326.

Sadin, Eric (2021). *L'Intelligence artificielle ou l'enjeu du siècle. Anatomie d'un antihumanisme radical*. Paris: L'Echappée.

49

WALKING ART

Antoine Freychet, in collaboration with Anastasia Chernigina

Whether following a precise itinerary or wandering around more freely, an art walk places us in relation to our surroundings and then consolidates and extends that relationship. The practice crosses disciplines such as choreography, sound studies, and cineplastics. Despite their distinctive features, different types of art walks share ecosophical traits, including the production of spontaneous and deliberate connections to place; the reconfiguration of ways of thinking about the world and our place within it; and the desire to share our emotions, to create an intimate, communal experience.

Critical Potential

As a form – and it is in this respect it can be described as "in transition" – a walk is positioned against the production mindset underpinning most artistic institutions and industries. Therefore, it can be viewed in line with degrowth movements in the sense that it only requires minimal resources and because it produces not a surplus, but a collection of displacements: displacements around the subject, collective ways of being, ways of experimenting with time and space (see Davila 2002). The form also calls into question established relationships between artists, viewers, and the designated context of creation. An art walk leads participants to create their own ways of experiencing a milieu, thereby understanding that they possess an aesthetic dimension themselves as a milieu, situation, experience, or process. The group of non-human actors that populate the environment fully participate in this process of artistic creation.

The art walk can be conceptualised as a conscious, performative challenge to the fragmentation of space, social issues, and inner emotions that attempts to resist the erosion of common and shared experiences (Biserna 2015: 26). It is

DOI: 10.4324/9781003455523-50

an "empowering" practice, which opens up new ways of experiencing the landscape that contradict limitations and hierarchies (spatial, social, emotional; see Biserna 2015: 34). An art walk emphasises our power to make, propose, create, and produce subjectivity and commonality. This commonality extends to non-humans through a concern for equally sharing the emotional world between different individuals and species.

Integration into Everyday Life

Integrating an art walk into everyday situations can destabilise what constitutes the habitual by highlighting the infraordinary. By opposing planning, optimisation, and profitability, the unpredictable and the provisional are found within the ordinary, making it possible to reintegrate it into the imaginary and the critical, and reinstating its vital importance (see Pérec 1989). This shift comes from paying attention to detail, to the marginal, or to what is normally imperceptible, but also by developing an awareness of the material, temporal, and relational complexity and fragility of all milieus. In all cases, the infraordinary requires a particular type of attention that is freed from usual preoccupations and solicitations.

Let's take the example of the sound that signals the closing of doors on a metro train. If the commuter pays any attention to it at all, it is generally only acknowledged for its informative content. In fact, there is a strange chance that this sound will be completely pushed out and rejected from the field of awareness. Its sonorous characteristics and relationships, as well as its affective dimension, are thus masked. Acknowledging that this example refers to an acoustically striking and prominent sound, it is not difficult to imagine what happens to other sounds (the creaking of seats, the clanking of handles, or even the rats talking about their day at the entrance to underground tunnels). Against this context of desensitised listening, reinforced by the progressive saturation of our soundscapes, sound walks reintroduce musicality into everyday life, and thus find their purpose, their disruptive strength (see Westerkamp 2006).

An Opportunity for Learning

In addition, art walks address the construction and transformation of our own subjectivity in relation to collectivity and place. A walk is always situated and dynamic; through it, the landscapes encountered and the events experienced acquire a unique coherence that gradually, and generally collectively, alters and enriches our ways of thinking and feeling. A walk is more than the opportunity for an aesthetic experience; it is also an experience of learning or, more precisely, the dynamics of learning. Here, two issues are in play: it facilitates the development of an ecological sensitivity (an ecologisation of emotion) and the production of knowledge through the emotions. In this respect, it is a re-evaluation of the role of emotions in our ways of understanding the world and improving our actions. These dynamics of learning address the self, the group, the places encountered, and the human, animal, and vegetation activities that take place there as well as how all those elements come together to create a milieu. These dynamics are charged with a motivation, a curiosity, and a specific strength that comes from experience. They can also reinforce our sense of belonging to and our involvement in the places we live.

The learning experience of a sound walk aims to develop different aptitudes (see Valentine 2003): (a) an awareness of soundscapes and their importance for our existence and for that of all living things since sound conveys affects and information from what surrounds us; (b) a lucidity and sensitivity towards the conditions in which sounds emerge and we listen to them, as well as our impact on their production (since a soundscape depends on those who live and experience it, it confers a responsibility on them); (c) an ability to both conceptualise and describe the sensations, meanings, and discoveries experienced; and (d) a disposition, drawing on the previous ones, to appreciate or implement processes of "artistic" realisation and actions to transform the soundscape (limiting certain sounds whilst welcoming or making others). Learning to inhabit the soundscape through active listening is therefore indistinguishable from the artistic dimension of a sound walk; indeed, it is part of it. The way of understanding the world that results from a sound walk directly impacts our ways of perceiving, feeling, and being moved.

Strategies

As a walk can engage people who are not or seldom used to the practices and issues that link arts and ecologies, choices must be made to adapt the planned routes to the listeners' needs. Amongst other possibilities, it is possible to present quite stark contrasts or even to choose places where activities are numerous or varied. Signposting to notable features or to maintain safety and respect for the environment and its inhabitants is also important. If necessary, the presence of habitants that we do not often think of (shrubs, insects, and other beings) can be highlighted. A walk is based on a set of directions that guide participants and allow them to disconnect from their normal sensory imperatives. How open the experience is or how much perceptive freedom participants are afforded depends on the strategy adopted. Being too heavily prescriptive can impose a specified emotional experience in a rather harsh way. In contrast, if there are too few instructions, the place offered to these emotions can be passed over.

Spending time to consider these questions is worthwhile: indeed, it is required to create the conditions to lead participants towards an emotional space that can be, consecutively or simultaneously, open, receptive, and active;

that can "produce" individual and shared subjectivities, as well as new and situated ecological questions, all taking place in minimal material conditions. Next time you go out for a walk, why not start listening to the rustlings of your body moving and the sounds that these movements provoke? Next, listen to the wind and the rain if there is any: what are they causing to make sound? Then, open your ears to the entire soundscape. Start by identifying the loudest sound, then the most high pitched or the closest. After that, extend your listening to identify all the sounds you can distinguish. Finally, look for the acoustic relationships between the sounds, such as between the rhythm and expansiveness of your breath and the road traffic. Search for them between the density and distance of the chorus of metro turnstiles and the escalators, between the acoustic space of cicadas singing and the creaking pine trees. If anyone asks what you are doing, immediately tell them it is a soundwalk and invite them to join in.

Three Key References

Biserna, Elena (2015). "Mediated Listening Paths: Breaking the Auditory Bubble". *Wi: Journal of Mobile Media* 9(2): 26–38. URL: https://issuu.com/jessicathompson.ca/docs/6_locussonus_proceedings

Perrin, Julie (2017). "Traverser la ville ininterrompue: sentir et se figurer à l'aveugle. À propos de *Walk, Hands, Eyes (a city)* de Myriam Lefkowitz". *Ambiances* 3. URL: http://ambiances.revues.org/962

Westerkamp, Hildegard (2006). "Soundwalking as Ecological Practice", in *The West Meets the East in Acoustic Ecology*, Keiko Torigoe, Tadahiko Imada, and Kozo Hiramatsu (eds) Hirosaki: Hirosaki University, 84–9.

BIBLIOGRAPHY

(L') Agence Touriste, Mathias Poisson and Virginie Thomas (2013). *Comment se perdre sur un GR*. Marseille: Editions Wildproject.

(L') Ecole Supérieure d'Arts Annecy Alpes, and Centre de la Photographie de Genève (2021). "Effondrement des Alpes". URL: https://www.esaaa.fr/eda/presentation/

Abram, David (1996). *The Spell of the Sensuous: Perception and Language in a More-Than-Human World*. New York: Pantheon Books.

——— (2013). *Comment la terre s'est tue. Pour une écologie des sens*. Paris: Les Empêcheurs de penser en rond/La Découverte.

Adorno, Theodor W. (1982). *Théorie Esthétique*, trans. Marc Jimenez. Paris: Klincksieck.

Afeissa, Stéphane-Hicham (ed) (2007). *L'éthique de l'environnement*. Paris: Vrin.

Agamben, Giorgio (1995). *Moyens sans fins. Notes sur la politique*. Paris: Rivages.

——— (2004). *The Open: Man and Animal*, trans. Kevin Attell. Stanford: Stanford University Press.

Agnew, Vanessa, Juliane Tomann, and Sabine Stach (eds) (2023). *Reenactment Case Studies: Global Perspectives on Experiential History*. London: Routledge.

Albarello, Luc, Jean-Marie Barbier, Étienne Bourgeois, and Marc Durand (eds) (2013). *Expérience, activité, apprentissage*. Paris: Presses Universitaires de France.

Albrecht, Glenn A. (2011). "Symbiocene". *Healthearth*, 19 May: http://healthearth. blogspot.com/2011/05/symbiocene.html (accessed 26 February 2022).

——— (2019). *Earth Emotions. New Words for a New World*. Ithaca and London: Cornell University Press.

Alexievitch, Svetlana ([1997] 1998). *La Supplication. Tchernobyl, chronique du monde après l'apocalypse*, trans. Galia Ackerman and Pierre Lorrain. Paris: J.-Cl. Lattès.

Anders, Günther ([1956 and 1980] 2002/2011). *Obsolescence de l'homme*, 2 vols. Paris: Éditions Encyclopédie des nuisances/Éditions Fario.

Anzaldúa, Gloria (1987). *Borderlands/La Frontera: The New Mestiza*. Portland: Aunt Lute Books.

———— (2009). *The Gloria Anzaldúa Reader*, AnaLouise Keating (ed) Durham, NC: Duke University Press.

Ardenne, Paul (2019). *Un art écologique. Création plasticienne et anthropocène.* Lormont: Le Bord de l'Eau.

Arecchi, Tito (2011). "Dinamica della cognizione. Complessità e creatività", in *Paesaggi della complessità. La trama delle cose e gli intrecci tra natura e cultura*, Roberto Barbanti, Luciano Boi, and Mario Neve (eds) Milan: Mimesis Edizioni, 55–88.

Artaud, Antonin (2000). "'Mise en scène' and Metaphysics", in *The Routledge Reader in Politics and Performance*. Jane de Gay and Lizbeth Goodman (eds) Abingdon: Routledge, 98–101.

Assouly, Olivier (2008). *Le Capitalisme esthétique.* Paris: CERF.

Astruc, Rémi (ed) (2015). *La Communauté revisitée/Community Redux.* Versailles: RKI Press.

Augendre, Marie, Jean-Pierre Llored, and Yann Nussaume (eds) (2018). *La Mésologie, un autre paradigme pour l'Anthropocène? Actes du colloque de Cerisy-la-Salle.* Paris: Hermann.

Augoyard, Jean-François, and Henry Torgue (eds) (1995). *À l'écoute de l'environnement. Répertoire des effets sonores.* Marseille: Parenthèses.

Azoulay, Ariella (2008). *The civil contract of photography.* New York: Zone Brooks.

Barbanti, Roberto (2009). *Les origines des arts multimedia.* Nîmes: Lucie éditions.

———— (2011). "Écologie sonore et technologies du son". *Sonorités* 6: 9–41.

———— (2016). "Du *Monde de l'art* à l'art dans le monde. Pour une décroissance de l'art?", in *Art i decreixement/Arte y decrecimiento/Art et décroissance*, Carmen Pardo (ed) Girone: Documenta Universitaria, 33–54.

———— (2017). *Chimere dell'arte. Guerra estetica, ultramedialità e arte genetica.* Verona: Ombre Corte.

Barbanti, Roberto, Silvia Bordini, and Lorraine Verner (2012). "Art, paradigme esthétique et écosophie". *Chimères* 6: 115–23.

Barbanti, Roberto (2013). "Art, paradigme esthétique et écosophie", in *Théories et pratiques écologiques. De l'écologie urbaine à l'imagination environnementale*, Manola Antonioli (ed) Paris: Presses Universitaires de Paris Ouest, 317–34.

Barbour, Karen (2019). "Backyard activisms: Site dance, permaculture and sustainability". *Choreographic Practices* 10(1): 113–25.

Barbour, Karen, Vicky Hunter, and Melanie Kloetzel (eds) (2019). *(Re)Positioning Site Dance: Local Acts, Global Perspectives.* Bristol and Chicago: Intellect Ltd.

Bardet, Marie, Joanne Clavel, and Isabelle Ginot (eds) (2019). *Écosomatiques. Penser l'écologie depuis le geste.* Montpellier: Editions Deuxième Époque.

Bar-On, Yinon M., Rob Phillips, and Ron Milo (2018). "The biomass distribution on earth". *Proceedings of the National Academy of Sciences* 115(25). URL: https://doi.org/10.1073/pnas.1711842115

Barthes, Jacques-Olivier (2012). "L'affichage électoral à la fin du XIXe siècle", in *Affiche-Action: quand la politique s'écrit dans la rue*, Béatrice Fraenkel, Magali Gouiran, Nathalie Jakobowicz, and Valérie Tesnière (eds) Paris: Gallimard, 94–9.

Baumgarten, Alexander G. (1988). *Esthétique précédée des Méditations philosophiques sur quelques sujets se rapportant à l'essence du poème et de la Métaphysique.* Paris: L'Herne.

Bazin, André ([1945] 1985). "Ontologie de l'image photographique", in *Qu'est-ce que le cinéma?.* Paris: Éditions du Cerf.

Beau, Rémi, and Catherine Larrère (eds) (2018). *Penser l'Anthropocène*. Paris: Presses de Science Po.

Beaufils, Eliane (2019a). "Déplacement de la scène de l'*anthropos* dans *Conversations déplacées* d'Ivana Müller". *thaêtre* 4. URL: https://www.thaetre.com/2019/06/01/conversations-deplacees/ (accessed 11 July 2023).

――― (2019b). "L'art participatif peut-il enfanter le citoyen-à-venir?". *Nectart* 9(2): 136–45.

――― (2022). "Staging Larger Scales and Deep Entanglements. The Choice of Immersion in Four Ecological Performances", in *Life, Re-Scaled: The Biological Imagination in Twenty-First-Century Literature and Performance*, Liliane Campos and Pierre-Louis Patoine (eds) Cambridge: Open Book Publishers, 353–78. URL: https://www.openbookpublishers.com/books/10.11647/obp.0303/chapters/10.11647/obp.0303.13

――― (2023). "Expériences de pensée écofictionnelles sur les scènes contemporaines: *Unlikely Creatures (drei) – us hearing voices* (2018). de Billinger & Schulz et *new skin* (2018). de Hannah de Meyer". *Fabula-LhT* 29. URL: http://www.fabula.org/lht/29/beaufils.html

Becattini, Giacomo (2015). *La coscienza dei luoghi. Il territorio come soggetto corale*. Roma: Donzelli.

Bénichou, Anne (ed) (2015). *Recréer/Scripter: mémoires et transmissions des œuvres performatives et chorégraphiques contemporaines*. Dijon: Presses du Réel.

Benjamin, Walter ([1916] 2000). "Sur le langage humain et sur le langage en général", in *Œuvres I*, trans. Maurice de Gandillac, Rainer Rochlitz, and Pierre Rusch. Paris: Gallimard, 142–65.

――― (2002a). *The Arcades Project*. Cambridge, MA: Harvard University Press.

――― (2002b). *Charles Baudelaire. Un poète lyrique à l'apogée du capitalisme*. Paris: Éditions Payot.

――― (2002c). "The Work of Art in the Age of Mechanical Reproduction", in *Walter Benjamin: Selected Writings. Volume 3: 1935–1938*, Howard Eiland and Michael W. Jennings (eds) Cambridge, MA and London: Belknap Press, 100–33.

――― (2003). *Walter Benjamin: Selected Writings. Volume 4: 1938–1940*, Howard Eiland and Michael W. Jennings (eds) Cambridge, MA and London: Belknap Press.

Bergala, Alain (1996). "Critique/théorie: l'évaluation et la preuve". *Cinemas* 6(2–3): 29–44.

Bergson, Henri (1946). *The Creative Mind*, trans. Mabelle L. Andison. New York: Philosophical Library.

Berque, Augustin (1987). *Écoumène. Introduction à l'étude des milieux humains*. Paris: Belin.

――― (1999). *Les raisons du paysage, de la Chine antique aux environnements de synthèse*. Paris: Editions Hazan.

――― (2000). *Médiance de milieu en paysages*. Paris: Belin.

――― (2014). *Poétique de la terre, essai de mésologie*. Paris: Belin.

――― (2016). *Écoumène. Introduction à l'étude des milieux humains*. Paris: Belin.

――― (2019). *Poetics of the Earth. Natural History and Human History*, trans. Anne-Marie Feenberg-Dibon. London: Routledge.

Besse, Jean-Marc (2003). *Face au monde. Atlas, jardins, géoramas*. Paris: Desclée de Brouwer.

Beuys, Joseph (1988). *Par la présente, je n'appartiens plus à l'art*. Paris: L'Arche Éditeur.

Bigé, Romain (2020). "Danser l'Anarchie: théories et pratiques anarchistes dans le Judson Dance Theater, Grand Union et le Contact Improvisation". *Revista Brasileira de Estudos Presença* 10(1). URL: http://dx.doi.org/10.1590/2237-266089064

Bihouix, Philippe (2014). *L'Âge des low tech: Vers une civilisation techniquement soutenable*. Paris: Seuil.

Biserna, Elena (2015). "Mediated Listening Paths: Breaking the Auditory Bubble". *Wi: Journal of Mobile Media* 9(2): 26–38. URL: https://issuu.com/jessicathompson.ca/docs/6_locussonus_proceedings

Bishop, Claire (2012). *Artificial Hells. Participatory Art and the Politics of Spectatorship*. London: Verso.

Blanc, Nathalie (2008). *Vers une esthétique environnementale*. Versailles: Quae.

Blanc, Nathalie, and Barbara Benish (2016). *Form, Art, and the Environment: Engaging in Sustainability*. London: Routledge.

Blanc, Nathalie, Denis Chartier, and Thomas Pughe (2008). "Littérature & écologie: vers une écopoétique". *Ecologie & Politique* 2(36): 15–28.

Blanc, Nathalie, and Jacques Lolive (eds) (2007). *Esthétique et espace public*, Cosmopolitiques 15. Rennes: Apogée.

Blanchot, Maurice ([1950] 1992). "La condition critique". *Trafic* 2: 140–2.

Blümlinger, Christa (2013). *Cinéma de seconde main. Esthétique du remploi dans l'art du film et des nouveaux médias*. Paris: Klincksieck.

Boccali, Renato (2019). "Sur l'intercorporéité et l'interanimalité. Merleau Ponty et la chair primordiale". *Revue de métaphysique et de morale* 101: 39–49.

Bogre, Michelle (2012). *Photography as activism, Images for Social Change*. New York: Social Press.

Bogue, Ronald (2009). "A Thousand Ecologies", in *Deleuze/Guattari & Ecology*, Bernd Herzogenrath (ed) New York: Palgrave MacMillan, 42–56.

Böhme, Gernot (2017). *The Aesthetics of Atmospheres*, Jean-Paul Thibaud (ed) London: Routledge.

Bolter, David Jay, and Richard Grusin (2000). *Remediation. Understanding New Media*. Cambridge, MA: MIT Press.

Bonneuil, Christophe, and Jean-Baptiste Fressoz (2013). *L'évenement Anthropocène*. Paris: Seuil.

Bookchin, Murray (1993). "What is social ecology?", in *Environmental Philosophy: From Animal Rights to Radical Ecology*, Michael E. Zimmerman (ed) Englewood Cliffs, NJ: Prentice Hall.

——— (2007). *From Social Ecology and Communalism*. Oakland: AK Press.

——— (2019). *Pouvoir de détruire, pouvoir de créer. Vers une écologie sociale et libertaire*. Paris: L'Échappée.

Bottiglieri, Carla (2014). "Médialités: quelques hypothèses sur les milieux de Feldenkrais", in *Penser le somatiques avec Feldenkrais. Politiques et esthétiques d'une pratique corporelle*, Isabelle Ginot (ed) Paris: L'Entretemps, 77–114.

Boumard, Patrick (1996). *Célestin Freinet*. Paris: Presses Universitaires de France.

Boumédiene, Samir (2019). *La colonisation du savoir. Une histoire des plantes médicinales du "Nouveau Monde" (1492–1750)*. Vaulx-en-Velin: Les Éditions des mondes à faire.

Bourriaud, Nicolas (2001). *Esthétique relationnelle*. Dijon: Les Presses du Réel.

——— (2021). *Inclusions. Esthétique du Capitalocène*. Paris: Presses Universitaires de France.

———— (2022). *Inclusions. Aesthetics of the Capitalocene*. London: Sternberg Press.

Bouvet, Rachel (2013). "Géopoétique, géocritique, écocritique: points communs et divergences". URL: https://rachelbouvet.files.wordpress.com/2013/05/confecc81rence_angers-28-mai-2013.pdf (accessed 14 July 2023).

Brereton, Pat (2005). *Hollywood Utopia: Ecology in Contemporary Cinéma*. Portland/Bristol: Intellect Books.

Breteau, Clara (2018). *POÈME: la POïesis à l'Ère de la Metamorphose*, PhD dissertation. University of Leeds, UK.

———— (2019). "POÈME: la POïesis à l'Ère de la MEtamorphose". *Ecozon@* 10(1): 136–63.

———— (2022). *Les vies autonomes, une enquête poétique*. Arles: Actes Sud.

Brondizio, Eduardo S., et al. (2016). "Re-conceptualizing the Anthropocene: A call for collaboration". *Global Environmental Change* 39: 318–27.

Brown, Andrew (2014). *Art and Ecology Now*. London: Thames & Hudson.

Brunet, François (2012). *L'Invention de l'idée de photographie*. Paris: PUF.

Buell, Lawrence (1995). *The Environmental Imagination: Thoreau, Nature Writing, and the Formation of American Culture*. Cambridge, MA and London: Belknap Press.

Buren, Daniel ([1970] 1973a). "Function of the Museum". *Artforum* 12(1): 68.

———— (1973b). "The Function of an Exhibition". *Studio International* 186(961): 216.

Buxton, William (2008). "My vision isn't my vision: Making a career out of getting back to where I started", in *HCI Remixed: Reflections on works that have influenced the HCI community*, Thomas Erickson and David McDonald (eds) Cambridge: MIT Press, 7–12.

Cage, John (1973). "45' for a speaker", in *Silence: Lectures and Writings*. Middletown, CT: Wesleyan University Press, 146–93.

———— (1976). *A Year from Monday*. London: Marion Boyars.

Caillet, Aline (2014). *Dispositifs critiques. Le documentaire, du cinéma aux arts visuels*. Rennes: Presses Universitaires de Rennes.

Caillet, Aline, and Frédéric Pouillaude (eds) (2017). *Un art documentaire. Enjeux esthétiques, politiques et éthiques*. Rennes: Presses Universitaires de Rennes.

Callicott, J. Baird, and Michael P. Nelson (eds) (1998). *The Great New Wilderness Debate*. Athens: University of Georgia Press.

Canetti, Elias (1962). *Crowds and Power*, trans. Carol Stewart. New York: Viking Press.

Canguilhem, Georges ([1952] 2008). *Knowledge of Life*, trans. Stefanos Geroulanos and Daniela Ginsburg. New York: Fordham University Press.

Jimenez Carmona, Susana, and Carmen Pardo Salgado (eds) (2021). "Aperturas y Derivas del Arte Sonoro". *Laocoonte. Revista de Estética y Teoría de las Artes* 8: 49–56. URL: https://ojs.uv.es/index.php/LAOCOONTE/issue/viewIssue/1486/313

Carson, Rachel ([1962] 2002). *Silent Spring*. Boston: Mariner Books.

Casemajor, Nathalie, and Will Straw (2017). "The Visuality of Scenes. Urban Cultures and Visual Scenescapes". *Imaginations. Journal of cross-cultural image studies* 7(2): 4–19.

Castoriadis, Cornelius ([1975] 1997). *The Imaginary Institution of Society*, trans. Kathleen Blamey. Cambridge: Polity Press.

Ceraso, Steph (2014). "(Re)Educating the Senses: Multimodal Listening, Bodily Learning, and the Composition of Sonic Experiences". *College English* 77(2): 102–23.

Chabert, Garance, and Aurélien Mole (2018). *Les artistes iconographes*. Dijon: Les Presses du Réel.

Chamoiseau, Patrick, and Michel Le Bris (eds) (2018). *Osons la fraternité. Les écrivains aux côtés des migrants*. Paris: Philippe Rey.

Chan, Kai M. A., et al. (2016). "Why protect nature? Rethinking values and the environment". *Proceedings of the National Academy of Sciences of the United States of America* 113(6): 1462–5.

Charles, Daniel (1998). *Musiques nomades*. Paris: Kimé.

Chateau, Dominique (1999). *Arts plastiques. Archéologie d'une notion*. Nîmes: Jacqueline Chambon.

Chéroux, Clément (2015). *Avant l'avant-garde, du jeu en photographie 1890–1940*. Paris: Textuel.

Chevrier, Jean-François, and Philippe Roussin (eds) (2001/2006). "Le parti pris du document: littérature, photographie, cinéma et architecture au XXᵉ siècle". *Communications*, 71 and 79.

Choay, Françoise (ed) (2004). "Introduction", in *L'Art d'édifier*, Leon Battista Alberti (ed) Paris: Seuil, 22–6.

——— (2011). *La terre qui meurt*. Paris: Fayard.

Citton, Yves (2013). "Économie de l'attention et nouvelles exploitations numériques". *Multitudes* 54(3): 165–75.

——— (ed) (2014). *L'Économie de l'attention. Nouvel horizon du capitalisme?*. Paris: La Découverte.

Citton, Yves, and Jacopo Rasmi (2020). *Générations collapsonautes. Perspectives d'effondrement*. Paris: Seuil.

Clavel, Joanne (2017). "Expériences de Natures, investir l'écosomatique", in *Le souci de la nature, Apprendre Inventer, Gouverner*, Cynthia Fleury and Anne-Caroline Prévot (eds) Paris: CNRS, 257–69.

Clavel, Joanne, Vincent Devictor, and Romain Julliard (2011). "Worldwide decline of specialist species: theory and application". *Frontiers in Ecology and the Environment* 9(4): 222–9.

Clavel, Joanne, and Isabelle Ginot (2015). "Pour une écologie des somatiques?". *Revista Brasileira de Estudos da Presença* 5(1): 85–100.

Clavel, Joanne, and Camille Noûs (2020). "Planetary Dance d'Anna Halprin, étoile d'une constellation kinesthésique écologique". *Technique et Culture* 74: 174–7.

Claver, Ainhoa Kaiero (2010). "Les concerts de cloches de Llorenç Barber et la conception postmoderne de l'espace urbain". *Filigrane. Musique, esthétique, sciences, société* 12. URL: https://revues.mshparisnord.fr/filigrane/index.php?id=299

Clément, Gilles (2004). *Manifeste du Tiers paysage*. Paris: Sujet/Objet.

——— (2012). *Jardins, paysage et génie naturel*. Paris: Collège de France/Fayard. URL: http://lecons-cdf.revues.org/510; DOI: 10.4000/lecons-cdf.496

——— ([2009] 2014). *L'Alternative ambiante*. Paris: Sens & Tonka.

——— (2015). *The Planetary Garden and Other Writings*. Philadelphia: University of Pennsylvania Press.

Cohendet, Patrick, David Grandadam, and Laurent Simon (2010). "The Anatomy of the Creative City". *Industry & Innovation* 17(1): 91–111.

Coles, Laura Lee (2021). "Landscape, eco-arts practice, and digital technology in the public realm", in *The Routledge Companion to Art in the Public Realm*, Cameron Cartiere and Leon Tan (eds) New York: Routledge.

Collins, Nicolas ([2006] 2020). *Handmade Electronic Music. The Art of Hardware Hacking*, third edition. New York: Routledge.

Colombo, Fabien, Nestor Engone Elloué, and Bertrand Guest (eds) (2018). *Écologie et Humanités*. Special Issue of *Essais, Revue interdisciplinaire d'Humanités* 13. URL: http://journals.openedition.org/essais/411 (accessed 1 December 2019).

Contini, Leone, and Cecilia Guida (2017). "What Will It Happen to Earth if Seeds go Crazy? A Conversation About the 'Bank of the Migrating Germplasm'". *Visual Ethnography* 6(2): 203–22.

Contour, Catherine (ed) (2017). *Une plongée avec Catherine Contour – Créer avec l'outil hypnotique*. Paris: Naïca éditions.

Cosgrove, Denis, and Stephen Daniels (1998). *The Iconography of Landscape: Essays on the Symbolic Representation, Design and Use of Past Environment*. Cambridge: Cambridge University Press.

Couchot, Edmond (2007). "Réinventer le temps à l'heure du numérique". *Interin* 4(2): 1–11.

Courtecuisse, Claude (2005). "Les déplacements immobiles", in *Un léger déplacement dans l'ordre des choses*. Bruxelles: Quo Vadis.

—— ([1994] 2011). "Les petits jardins clos d'herbe errante", in *La ville fertile, vers une nature urbaine*. Special Issue of *Paysages Actualités* (March): 55.

Crary, Jonathan (1990). *Techniques of the Observer. On Vision and Modernity in the Nineteenth Century*. Cambridge, MA: MIT Press.

—— (2016). *24/7: Le capitalisme à l'assaut du sommeil*. Paris: La Découverte.

Crawford, Kate, and Vladan Joler (2018). "Anatomy of an AI System". URL: https://anatomyof.ai/ (accessed 11 January 2021).

Crawford, Matthew B. (2014). *The World Beyond Your Head. On Becoming an Individual in an Age of Distraction*. New York: Farrar, Straus and Giroux.

Crimp, Douglas (1986). "Serra's Public Sculpture: Redefining Site Specificity", in *Richard Serra Sculpture*, Rosalind Krauss (ed) New York: The Museum of Modern Art, 40–56.

Criton, Pascale (2016). "L'écoute plurielle", in *Musique et écologies du son. Propositions théoriques pour une écoute du monde*, Makis Solomos, Roberto Barbanti, Guillaume Loizillon, Kostas Papparrigopoulos, and Carmen Pardo (eds) Paris: L'Harmattan, 19–33.

Cronon, William (1995). "The trouble with wilderness; or, getting back to the wrong nature", in *Uncommon Ground. Rethinking the Human Place in Nature*, William Cronon (ed) New York: W. W. Norton & Co., 69–90.

Crutzen, Paul J. (2007). "La géologie de l'humanité: l'Anthropocène". *Écologie & Politique* 34(1): 141–8.

Cubitt, Sean, Salma Monani, and Stephen Rust (eds) (2013). *Ecocinema Theory and Practice*. New York: Routledge.

Cukierman, Leïla, Gerty Dambury, and Françoise Vergès (eds) (2018). *Décolonisons les arts!*. Paris: L'Arche.

d' Eaubonne, Françoise (1974). *Le Féminisme ou la mort*. Paris: Horay.

D'Alisa, Giacomo, Federico Demaria, and Giorgos Kallis (eds) (2015). *Degrowth. A Vocabulary for a New Era*. London: Routledge.

Daniele, Daniela (1997). "Corpi imprevisti e donne in performance", in *Meduse cyborg. Antologia di donne arrabbiate*. Milan: Shake, 7–11.

Danowski, Déborah, and Eduardo Viveiros De Castro (2014). "L'arrêt du monde", in *De l'univers clos au monde infini*, Émilie Hache (ed) Bellevaux: Éditions Dehors, 221–339.

Dardot, Pierre, and Christian Laval (2014). *Commun. Essai sur la révolution au XXe siècle*. Paris: La Découverte.

Davila, Thierry (2002). *Marcher, créer: déplacements, flâneries, dérives dans l'art de la fin du XXe siècle*. Paris: Éditions du Regard.

Debord, Guy Ernest (1989). "Rapport sur la construction des situations et sur les conditions de l'organisation et de l'action de la tendance situationniste internationale". *Inter* 44: 1–11.

Deleuze, Giles (2012). *Gilles Deleuze from A to Z*, trans. Charles J. Stivale. Cambridge, MA: MIT Press/Semiotext(e). URL: https://deleuze.cla.purdue.edu/seminars/gilles-deleuze-abc-primer/lecture-recording-3-n-z

Deleuze, Gilles, and Félix Guattari (1980). *Mille plateaux*. Paris: Les éditions de Minuit.

Deleuze, Gilles (1987). *A Thousand Plateaus: Capitalism and Schizophrenia*, trans. Brian Massumi. Minneapolis: University of Minnesota Press.

Delon, Gaspard, Charlie Hewison, and Aymeric Pantet (eds) (2023). *Écocritiques. Cinéma, Audiovisuel, Arts*. Paris: Hermann.

Delorme, Damien (2019). "Poétiser la transition écologique". *Les Cahiers de la Justice* 3: 537–49.

Delorme, Stéphane (2019). "Cahiers Verts". *Cahiers du Cinéma* 754: 5.

Delphy, Christine ([1991] 2001). "Penser le genre. Problèmes et résistances", in *L'ennemi principal 2: Penser le genre*. Paris: Syllepse, 243–60.

Demarco, Richard (1982). "Conversations with Artists". *Studio International* 195(996): 46.

Demos, T. J. (2016). *Decolonizing Nature: Contemporary Art and the Politics of Ecology*. Berlin: Sternberg Press.

Deng, Shanshan, Suxia Liu, Xingguo Mo, Liguang Jiang, and Peter Bauer-Gottwein (2021). "Polar Drift in the 1990s Explained by Terrestrial Water Storage Changes". *Geophysical Research Letters* 48(7). URL: https://agupubs.onlinelibrary.wiley.com/doi/epdf/10.1029/2020GL092114

Deriu, Marco (2015). "Autonomy", in *Degrowth. A Vocabulary for a New Era*, Giacomo D'Alisa, Federico Demaria, and Giorgos Kallis (eds) London: Routledge, 55–8.

Desanti, Jean-Toussaint (1983). "L'obscène ou les malices du signifiant". *Traverses* 29: 128–33.

Descartes, René (2006). *A Discourse on the Method of Correctly Conducting One's Reason and Seeking Truth in the Sciences*. New York: Oxford University Press.

Descola, Philippe (2005). *Par-delà nature et culture*. Paris: Gallimard.

Despret, Vinciane, and Michel Meuret (2016). "Cosmoecological sheep and the arts of living on a damaged planet". *Environmental Humanities* 8(1): 24–36.

Dewey, John (1934a). *L'art comme expérience*. Paris: Gallimard.

——— (1934b). *La formation des valeurs*, trans. Alexandra Bidet, Louis Quéré, and Gérôme Truc. Paris: La Découverte.

——— (1934c). *Art as Experience*. New York: Capricorn Books.

——— (1980). *Art as Experience*. New York: Perigee Books.

——— (2014). *L'art comme expérience*. Paris: Gallimard.

Dhaliwal, Ranjodh Singh (2020). Online roundtable "Cities and Computers: Our Urban-Machinic Imaginaries". Online roundtable, December 4, UC Davis Humanities Institute, Davis, CA. URL: https://dhi.ucdavis.edu/events/cities-and-computers-our-urban-machinic-imaginaries

——— (2022). "On Addressability, or What Even is Computation?". *Critical Inquiry* 49(1): 1–27.

Didi-Huberman, Georges (1992). *Ce que nous voyons, ce qui nous regarde*. Paris: Les éditions de Minuit.

Dietze, Lena (2000). "Learning Is Living Acoustic Ecology as Pedagogical Ground. A Report on Experience". *The Journal of Acoustic Ecology* 1: 20–2.

DiMaggio, Paul J., and Walter W. Powell (1983). "The Iron Cage Revisited: Institutional Isomorphism and Collective Rationality in Organizational Fields". *American Sociological Review*, 48: 147–60.

DiScipio, Agostino (2013). *Pensare le tecnologie del suono e della musica*. Napoli: Editoriale Scientifica.

——— (2016). "Objet sonore? Événement sonore! Idéologies du son et biopolitique de la musique", in *Musique et écologies du son. Propositions théoriques pour une écoute du monde*, Makis Solomos, Roberto Barbanti, Guillaume Loizillon, Kostas Paparrigopoulos, and Carmen Pardo (eds) Paris: L'Harmattan, 35–46.

——— (2017). "Pratiques musicales, technologies contemporaines, économie de moyens. Quelques remarques générales", in *Transitions des arts, transitions esthétiques: processus de subjectivation et des-croissances*, Roberto Barbanti, Kostas Paparrigopoulos, Carmen Pardo, and Makis Solomos (eds) Paris: L'Harmattan, 57–74.

——— (2021). *Circuiti del tempo. Un percorso storico-critico nella creatività elettroacustica e informatico-musicale*. Lucca: LIM.

Dissanayake, Ellen (2010). *Homo Æstheticus. Where Art Comes From and Why*. Seattle: University of Washington Press.

Djebar, Assia (2006). *Discours de réception à l'Académie française*. URL: http://www. academie-francaise.fr/discours-de-reception-et-reponse-de-pierre-jean-remy

Dorlin, Elsa (2022). *Self Defense: A Philosophy of Violence*. New York: Verso.

Douglas, Mary (1966). *Purity and Danger. An Analysis of the Concepts of Pollution and Taboo*. London: Routledge.

Drever, John L. (2009). "Soundwalking: Aural Excursions into the Everyday", in *The Ashgate Research Companion to Experimental Music*, James Saunders (ed) Farnham: Ashgate, 163–92.

Duhautpas, Frédérick, and Makis Solomos (2019). "His master's voice (H. Westerkamp). Écoféminismes sonores", in *Compositrices, l'égalité en acte*, Laure Marcel-Berlioz, Omer Corlaix, and Bastin Gallet (eds) Paris: Éditions MF.

Duhem, Ludovic, and Richard Pereira de Moura (2020). *Design des territoires. L'enseignement de la biorégion*. Paris: Eterotopia.

Duncan, Isadora ([1928] 1998). *Ma vie*. Paris: Gallimard.

Duperrex, Matthieu (2018). Arcadies altérées. Territoires de l'enquête et vocation de l'art en Anthropocène, PhD dissertation. Université de Toulouse, France.

Dupont, Florence (1994). *L'Invention de la littérature. De l'ivresse grecque au texte latin*. Paris: La Découverte.

Durafour, Jean-Michel (2017). "Des extraterrestres aux manettes des images. Images, minéraux et cristaux à partir de *La Cité pétrifiée* (1957) de John Sherwood". *Images Re-vues* 14. URL: https://hal.science/hal-01666336/

Dussel, Enrique (2009). "Pour un dialogue mondial entre traditions philosophiques". *Cahiers des Amériques latines* 62: 111–27. URL: https://journals.openedition.org/cal/1619?lang=en

Edwards, Elizabeth (1992). *Photography and Anthropology 1860–1920*. New Haven: Yale University Press.

Eiermann, André (2009). *Postspektakuläres Theater. Die Alterität der Aufführung und die Entgrenzung der Künste*. Bielefeld: Transcript.

Esclapez, Christine (ed) (2014). *Ontologies de la création en musique, vol. 3: Des lieux en musique*. Paris: L'Harmattan.

Escobar, Arturo (2014). *Sentipensar con la tierra: Nuevas lecturas sobre desarrollo, territorio y diferencia*. Medellín: UNAULA.

Fagnart, Claire (2017). *La critique d'art*. Saint-Denis: Presses Universitaires de Vincennes.

Federici, Silvia (2004). *Caliban and the Witch: Women, the Body and Primitive Accumulation*. Brooklyn: Autonomedia.

Feld, Steven (1982). *Sound and Sentiment. Birds, Weeping, Poetics, and Song in Kaluli Expression*. Philadelphia: University of Pennsylvania Press.

Ferdinand, Malcom (2019). *Une écologie décoloniale. Penser l'écologie depuis le monde caribéen*. Paris: Seuil.

Ferrara, Patricia, and Bianca Millon-Devigne (eds) (2014). *Gestes de Terre et la Danse*. Toulouse: Groupe Unber Humber.

Ferreboeuf, Hugues (2018). *Lean ICT – Pour une sobriété numérique*. Rapport du groupe de travail sur le numérique pour le Shift Project. URL: https://theshiftproject. org/wp-content/uploads/2018/11/Rapport-final-v8-WEB.pdf (accessed 27 January 2023).

Ferreboeuf, Hugues, Maxime Efoui-Hess, and Xavier Verne (2021). "Impact environnemental du numérique: tendances à 5 ans et gouvernance de la 5G". URL: https:// theshiftproject.org/wp-content/uploads/2021/03/Note-danalyse_Numerique-et-5G_30-mars-2021.pdf (accessed 27 January 2023).

Ferreri, Mara (2015). "The seductions of temporary urbanism". *Ephemera* 15(1): 181–91.

Finburgh, Clare, and Carl Lavery (eds) (2015). *Rethinking the Theatre of the Absurd: Ecology, Environment and the Greening of the Modern Stage*. London: Bloomsbury.

Fischer-Lichte, Erika ([2004] 2008). *The Transformative Power of Performance: A New Aesthetics*. London: Taylor & Francis.

Flanagan, Richard (2020). "Australia Is Committing Climate Suicide". *The New York Times*, 3 January.

Fleury, Cynthia (2015). *Les Irremplaçables*. Paris: Gallimard.

Florida, Richard (2005). *Cities and the Creative Class*. London: Routledge.

Forti, Simone ([2003] 2009). *Oh, Tongue*, trans. Christophe Marchand Kiss. Genève: Al Dante.

Foster, Hal (1996). *The Return of the Real. The Avant-Garde at the End of the Century*. Cambridge, MA and London: MIT Press.

Foucault, Michel (1982). "The subject and power", in *Michel Foucault: Beyond Structuralism and Hermeneutics*, Hubert L. Dreyfus and Paul Rabinow (eds) Chicago: University of Chicago Press, 208–26.

———— (1984). *Histoire de la sexualité II. L'usage des plaisirs*. Paris: Gallimard.

Fraenkel, Béatrice, Magali Gouiran, Nathalie Jakobowicz, and Valérie Tesnière (eds) (2012). *Affiche-Action: quand la politique s'écrit dans la rue*. Paris: Gallimard.

Franco, Susanne, and Marina Nordera (eds) (2010). *Ricordanze: Memoria in movimento e coreografie della storia*. Milan: UTET Università.

Franko, Mark (2017). *The Oxford Handbook of Dance and Reenactment*. Oxford: Oxford University Press.

Fréchuret, Maurice (2019). *L'art et la vie*. Dijon: Les Presses du Réel.

Freire, Paulo (1970). *Pedagogy of the Oppressed*. New York: Continuum.

Fuchs, Elinor, and Una Chaudhuri (eds) (2002). *Land/Scape/Theater*. Ann Arbor: University of Michigan Press.

Galeota-Wozny, Nancy (2006). "Bird Brain Dance: entretien avec Jennifer Monson". *Nouvelles de Danse* 53: 212–36.

Gautel, Jakob (1999). *Sous le ciel de Paris*. Limoges: Sixtus Éditions.

George, Pierre (1993). "Crépuscule de l'homme habitant?". *Revue de géographie de Lyon* 4: 213–14.

Georgescu-Roegen, Nicholas (2011). *From Bioeconomics to Degrowth: Georgescu-Roegen's "New Economics" in Eight Essays*, Mauro Bonaiuti (ed) London: Routledge.

Ginot, Isabelle (1999). *Dominique Bagouet, un labyrinthe dansé*. Pantin: CND.

—— (eds) (2014). *Penser les somatiques avec Feldenkrais*. Lavérune: Éditions L'Entretemps.

Gintz, Claude (1992). "Vito Acconci, l'impossibilité de l'art public". *Art Press* 166: 10–18.

Glissant, Édouard (1990). *Poétique de la relation. Poétique III*. Paris: Gallimard.

Glissant, Édouard, and Patrick Chamoiseau (2009). *L'intraitable beauté du monde. Adresse à Barack Obama*. Paris: Galaade.

Glotfelty, Cheryll, and Harold Fromm (eds) (1996). *The Ecocriticism Reader: Landmarks in Literary Ecology*. Athens and London: University of Georgia Press.

Glotfelty, Cheryll, and Eve Quesnel (eds) (2015). *The Biosphere and the Bioregion. Essential Writings of Peter Berg*. New York: Routledge.

Godelier, Maurice (1984). *L'idéel et le matériel: pensée, économies, sociétés*. Paris: Fayard.

Goldberg, RoseLee (2018). *Performance Now: Live Art for the Twenty-First Century*. London: Thames & Hudson.

Goody, Jack (1977). *The Domestication of the Savage Mind*. Cambridge: Cambridge University Press.

Gorz, André (1967). *Strategy for Labor: A Radical Proposal*, trans. Martin A. Nicolaus and Victoria Ortiz. Boston: Beacon Press.

—— (1988). *Métamorphoses du travail quête de sens*. Paris: Éditions Galilée.

—— (2019). *Penser l'avenir. Entretien avec François Noudelmann*. Paris: La Découverte.

Graeber, David (2008). "The sadness of post-workerism, or 'Art and immaterial labour' conference, a sort of review". *The Commoner*, 1 April.

—— (2015). *The Utopia of Rules: On Technology, Stupidity, and the Secret Joys of Bureaucracy*. New York: Melville House Publishing.

Gremillet, David (2019). *Daniel Pauly, un océan de combat*. Marseille: Editions Wildproject.

Groener, Fernando, Rose-Maria Kandler, and Joseph Beuys (1987). *7000 Eichen, Joseph Beuys*. Cologne: Walther König.

Guattari, Félix (1989). *Les Trois Écologies*. Paris: Galilée.

—— (1995). *Chaosmosis: An Ethico-aesthetic Paradigm*, trans. P. Bains and J. Pefanis. Bloomington and Indiana: Indiana University Press.

—— (2000). *The Three Ecologies*, trans. Ian Pindar and Paul Sutton. London and New Brunswick, NJ: The Athlone Press.

—— (2011). "On contemporary art: Interview with Oliver Zahm, April 1992", in *The Guattari Effect*, Eric Alliez and Andrew Goffey (eds) London: Bloomsbury, 40–53.

—— (2013). *Qu'est-ce que l'écosophie?*. Paris: Lignes/Imec.

——— (2018). *Ecosophy*. London: Bloomsbury.

Guattari, Félix, and Suely Rolnik (2007). *Micropolitiques*. Paris: Les Empêcheurs de penser en rond.

Guénoun, Denis (2005). "Actions et adresses", in *Actions et acteurs. Raisons du drame sur scène*. Paris: Bellin, 80–1.

Guilbert, Laure (2011). *Danser avec le IIIème Reich. Les danseurs modernes et le nazisme*. Bruxelles: Éditions André Versailles.

Guillebaud, Christine (ed) (2017). *Toward an Anthropology of Ambient Sound*. London: Routledge.

Le Guin, Ursula K. (1989). "The carrier bag theory of fiction", in *Dancing at the Edge of the World*. New York: Grove Press, 165–71.

Guinard, Pauline (2016). "De la Peur et du géographe à Johannesburg: retour sur des expériences de terrain et propositions pour une géographie des émotions". *Géographie et cultures* 93–4: 277–301.

Gunkel, David J. (2015). *Of Remixology. Ethics and aesthetics after remix*. Cambridge, MA: MIT Press.

Hache, Emilie (ed) (2016). *Reclaim. Anthologie de textes ecoféministes*. Paris: Cambourakis.

Hage, Ghassan (2017). *Is Racism an Environmental Threat?*. Bristol: Polity Press.

Hallmann, Casper A., Martin Sorg, Eelke Jongejans, Henk Siepel, Nick Hofland, and Heinz Schwan (2017). "More than 75 percent decline over 27 years in total flying insect biomass in protected areas". *PLOS ONE* 12(10). URL: https://journals.plos.org/plosone/article?id=10.1371/journal.pone.0185809

Halprin, Anna (2009). *Mouvements de vie*, trans. Élise Argaud and Denise Luccioni. Bruxelles: Contredanse.

Hanse, Olivier (2010). *Utopies communautaires allemandes autour de 1900*. Saint-Étienne: Université Saint-Étienne.

Haraway, Donna (1988). "Situated Knowledges: The Science Question in Feminism and the Privilege of Partial Perspective". *Feminist Studies* 14(3): 575–99.

——— (1997). *Modest Witness@Second_Millenium.FemaleMan©_Meets_OncoMouse™: Feminism and Technoscience*. New York and London: Routledge.

——— (2003). *The Companion Species Manifesto. Dogs, People, and Significant Otherness*. Chicago: Prickly Paradigm Press.

——— (2016). *Staying with the Trouble*. Durham and London: Duke University Press.

——— (2020). *Vivre avec le trouble*, trans. Vivien Garcia. Vaulx-en-Velin: Edition des mondes à faire.

Hardin, Garrett (1968). "The Tragedy of the Commons". *Science* 162(3859): 1243–8.

Harman, Graham (2018). *Object-Oriented Ontology: A New Theory of Everything*. Toronto: Pelican.

Harney, Stefano, and Fred Moten (2013). *The Undercommons: Black Study and Fugitive Planning*. New York: Minor Compositions.

Harney, Stefano (2021). *All Incomplete*. Wivenhoe: Minor Composition.

Heidegger, Martin (1958). "La question de la technique", in *Essais et conférences*, trans. André Préau. Paris: Gallimard.

Helguera, Pablo (2011). *Education for Socially Engaged Art: A Materials and Techniques Handbook*. New York: Jorge Pinto Books.

Hetzel, Andreas V. (2013). "Situationen. Philosophische und künstlerische Annäherungen", in *Heterotopien. Perspektiven der intermedialen Ästhetik*, Nadja

Elia-Borer, Constanze Schellow, Nina Schimmel, and Bettina Wodianka (eds) Bielefeld: Transcript, 487–500.

Higgins, Dick (1966). "Intermedia". *The Something Else Newsletter* 1: 1–3.

Hince, Bernadette, Rupert Summerson, and Arnan Wiesel (eds) (2015). *Antarctica: Mmusic, Sounds and Cultural Connections*. Canberra: Australian National University Press.

Holland, Simon, Katie Wilkie, Paul Mulholland, and Allan Seago (eds) (2013). *Music and Human-Computer Interaction*. London: Springer.

Hopkins, Rob (2013). *The Power of Just Doing Stuff*. Cambridge: UIT/Green Books.

Hörl, Erich (2013). "A thousand ecologies. The process of cyberneticization and general ecology", in *The Whole Earth. California and the Disappearance of the Outside*, Diedrich Diederichsen (ed) Berlin: Sternberg: 121–30.

——— (2015). "The technological condition". *Parrhesia* 22: 1–15.

Hunter, Victoria (2017). "Experiencing space: Some implications for site-specific dance performance", in *Contemporary Choreography: A Critical Reader*, Jo Butterworth and Liesbeth Wildschut (eds) London: Routledge.

Huyghe, Pierre, and Éric Troncy (2014). "Systèmes sous-déterminés", in *Stream 03, Habiter l'Anthropocène*, Philippe Chiambaretta (ed) Paris: PCA, 125–41.

Illich, Ivan (2004a). *La Perte des sens*. Paris: Fayard.

——— (2004b). "Passé scopique et éthique du regard. Plaidoyer pour l'étude historique de la perception oculaire", in *La Perte des sens*. Paris: Fayard, 287–326.

Ingram, David (2000). *Green Screen: Environmentalism and Hollywood Cinema*. Exeter: University of Exeter Press.

Jaccottet, Philippe (1968). *L'Entretien des muses*. Paris: Gallimard.

——— (2014). *Œuvres*, José-Flore Tappy (ed) Paris: Gallimard.

Jacobs, Jane (1970). *The Economy of Cities*. New York: Vintage Books.

James, Christopher (2015). *The Book of Alternative Photographic Processes*. Boston: Cengage Learning.

Jordan, Stephanie (ed) (2000). *Preservation Politics: Dance Revived, Reconstructed, Remade*. London: Dance Books.

Juslin, Patrick N., and John A. Sloboda (eds) (2011). *Handbook of Music and Emotion. Theory, Research, Applications*. Oxford: Oxford University Press.

Kääpä, Pietari (2014). *Ecology and Contemporary Nordic Cinemas: From Nation-Building to Ecocosmopolitanism*. London/Oxford, New York: Bloomsbury.

Keislar, Douglas (2009). "A historical view to computer music technology", in *The Oxford Handbook of Computer Music*, Roger T. Dean (ed) Oxford: Oxford University Press, 11–43.

Kerzerho, Anne (2020). *L'entretien en danseur (à partir de l'expérience d'Autour de la table)*, MA dissertation. Université Paris 8, France.

Kester, Grant H. (2004). *Conversation Pieces: Community and Communication in Modern Art*. Berkeley: University of California Press.

——— (2011). *The One and the Many. Contemporary Collaborative Art in a Global Context*. Durham and London: Duke University Press.

Kloetzel, Mélanie (2019). "Site specific dance and environmental ethics: relational fields in the Anthropocene", in *(Re)Positioning Site Dance: Local Acts, Global Perspectives*, Karen Barbour, Vicky Hunter, and Melanie Kloetzel (eds) Bristol and Chicago: Intellect Ltd, 217–46.

Kostelanetz, Richard (ed) (1988). "Esthetics", in *Conversing with Cage*. New York: Limelight Editions.

Kramer, Paula, (2012). "Bodies, rivers, rocks and trees: Meeting agentic materiality in contemporary outdoor dance practices". *Performance Research* 17(4): 83–91.

Krause, Bernie (1987). "Bioacoustics: Habitat ambience and ecological balance". *Whole Earth Review* 57: 14–16.

——— (2015). *Voices of the Wild. Animal Songs, Human Din, and the Call to Save Natural Soundscapes*. New Haven and London: Yale University Press.

Krauss, Rosalind (1990). *Le Photographique – Pour une théorie des écarts*. Paris: Macula.

Kusama, Yayoi (2011). *Infinity Net: The Autobiography of Yayoi Kusama*. London: Tate Publishing.

Kwon, Miwon (2002). *One Place After Another. Site Specific Art and Locational Identity*. Cambridge, MA and London: MIT Press.

LaBelle, Brandon (2010). *Acoustic Territories/Sound Culture and Everyday Life*. New York: Continuum.

Lallement, Michel (2015). *L'âge du faire. Hacking, travail, anarchie*. Paris: Seuil.

Lamarre, Thomas (2012). "Humans and Machines". *Inflexions* 5: 29–67.

Landry, Charles (2008). *The Creative City: A Toolkit for Urban Innovation*. London: Earthscan.

LarbitsSisters (2018). "BitSoil POPup Tax and Hack Campaign". URL: https://www.larbitslab.be/bitsoil-popup-tax-hack-campaign/ (accessed 17 July 2023).

Larrère, Catherine (2016). "Nul ne sait ce que peut un environnement", in *Les limites du vivant. À la lisière de l'art, de la philosophie et des sciences de la nature*, Roberto Barbanti and Lorraine Verner (eds) Bellevaux: Éditions Dehors, 331–44.

Larrère, Catherine, and Raphaël Larrère (2015). *Du bon usage de la nature. Pour une philosophie de l'environnement*. Paris: Flammarion.

Larrère, Catherine (2015). *Penser et agir avec la nature. Une enquête philosophique*. Paris: La Découverte.

Latouche, Serge (2006). *Le pari de la décroissance*. Paris: Fayard.

Latour, Bruno ([1991] 1997). *Nous n'avons jamais été moderne. Essai d'Anthropologie symétrique*, second edition. Paris: La Découverte.

——— (2004). "Why Has Critique Run out of Steam? From matters of fact to matters of concern". *Critical Inquiry* 30(2): 225–48. URL: http://www.bruno-latour.fr/sites/default/files/89-CRITICAL-INQUIRY-GB.pdf

——— (2013). *An Inquiry Into Modes of Existence*, trans. Catherine Porter. Cambridge, MA and London: Harvard University Press.

——— (2015). *Face à Gaïa: huit conférences, sur le nouveau régime climatique*. Paris: La Découverte.

——— (2017). *Où atterrir? Comment s'orienter en politique*. Paris: La Découverte.

——— ([1994] 2018). "Esquisse d'un parlement des choses". *Écologie & Politique* 56(1): 47–64.

Launay, Isabelle (1996). *À la recherche d'une danse moderne: Rudolf Laban, Mary Wigman*. Paris: Chiron.

——— (ed) (2007). *Les Carnets bagouet, la passe d'une œuvre*. Dijon: Les Solitaires Intempestifs.

——— (2017). *Les danses d'après I. Poétiques et politiques des répertoires*. Pantin: CND.

——— (2019). *Les danses d'après II. Cultures de l'oubli et citations*. Pantin: CND.

Laville, Jean-Louis, and Antonio David Cattani (eds) (2006). *Dictionnaire de l'autre économie*. Paris: Gallimard.

Lehmann, Hans-Thies ([1998] 2002). *Le Théâtre postdramatique*, trans. Philippe-Henri Ledru. Paris: L'Arche.

Lejeune, Antoine, and Michel Delage (2017). *La mémoire sans souvenir*. Neuilly-sur-Seine: Odile Jacob.

Leopold, Aldo ([1949] 1995). *Almanach d'un conté des sables*. Paris: Aubier.

Lepecki, André (2016). "The body as archive, will to reenact and the afterlives of dances", in *Singularities: Dance in the Age of Performance*. London: Routledge.

Lessing, Wolfgang (2014). "Versuch über Technik", in *Verkörperungen der Musik. Interdisziplinäre Betrachtungen*, Jörn Peter Hiekel and Wolfgang Lessing (eds) Bielefeld: Transcript Verlag.

Lévi-Strauss, Claude (1966). *The Savage Mind*. Chicago: University of Chicago Press.

Lévy, Jacques, and Michel Lussault (eds) (2003). *Dictionnaire de la géographie et de l'espace des sociétés*. Paris: Belin.

Licht, Alan (2007). *Sound Art. Beyond Music, Between Categories*. New York: Rizzoli.

Lindell, John (1994). "On Site Specificity". *Documents* 4/5: 11–22.

Lipovetsky, Gilles, and Jean Serroy (2013). *L'esthétisation du monde. Vivre à l'âge du capitalisme artiste*. Paris: Gallimard.

Locatelli, Axelle (2019). *Les chœurs de mouvements Laban: entre danses traditionnelles, gymnastique et expériences artistiques (Allemagne, 1923–1936)*, PhD dissertation. Université Paris 8, France.

Locher, Fabien (2013). "Les pâturages de la Guerre froide: Garrett Hardin et la 'Tragédie des communs'". *Revue d'histoire moderne et contemporaine* 60(1): 3–36. URL: https://www.cairn.info/revue-d-histoire-moderne-et-contemporaine-2013-1-page-7.htm

Lu, Sheldon H., and Jiayan Mi (eds) (2009). *Chinese Ecocinema. In the Age of Environmental Challenge*. Hong Kong: Hong Kong University Press.

Lucier, Alvin (1995). *Reflections. Interviews, Scores, Writings*. Köln: MusikTexte.

Lugon, Olivier (2017). *Le Style documentaire: d'August Sander à Walker Evans, 1920–1945*. Paris: Editions Macula.

Lussier, Mark (2011). "Blake, Deleuze, and the emergence of ecological consciousness", in *Ecocritical Theory: New European Approaches*, Axel Goodbody and Kate Kigby (eds) Charlottesville and London: University of Virginia Press.

Lyotard, Jean-François (1981). "La performance et la phrase chez Daniel Buren", in *Performance Text(e)s & Documents*, Chantal Pontbriand (ed) Montréal: Parachute, 66–9.

———— (2013). *Why Philosophize?*, trans. Andrew Brown. Cambridge: Polity Press.

Mâche, François Bernard (1983). *Musique, mythe, nature, ou les dauphins d'Arion*. Paris: Klincksieck.

Madanipour, Ali (2018). "Temporary use of space: Urban processes between flexibility, opportunity and precarity". *Urban Studies* 55(5): 1093–110.

Madlener, Frank (2022). "Intelligence artificielle et imaginaire artistique". *Analyse Opinion Critique*. URL: https://aoc.media/opinion/2022/11/30/intelligence-artificielle-et-imaginaire-artistique/ (accessed 1 December 2022).

Maeder, Marcus (ed) (2010). *Milieux Sonores/Klangliche Milieus. Klang, Raum und Virtualität*. Bielefeld: Transcript.

Magnaghi, Alberto (2005). *The Urban Village. A Charter for Democracy and Local Self-Sustainable Development*. London: Zed Books.

———— (2014). *La biorégion urbaine, Petit traité du territoire bien commun*. Paris: Eterotopia.

———— (2017). *La conscience du lieu*. Paris: Eterotopia.

———— (2020). *Il principio territoriale*. Torino: Bollati Boringhieri.

Malaquais, Dominique (ed) (2019). *Kinshasa Chroniques.* Lyon: Descours.

Manovich, Lev (1998). "Navigable Space". URL: http://manovich.net/index.php/projects/navigable-space

Marchart, Oliver (2019). *Conflictual Aesthetics. Artistic Activism and the Public Sphere.* Berlin: Sternberg Press.

Marczewska, Kaja (2018). *This is not a copy. Writing at the iterative turn.* London: Bloomsbury.

Maris, Virginie (2010). *Philosophie de la biodiversité, Petite éthique pour une nature en péril.* Paris: Buchet-Chastel.

———— (2014). *Nature à vendre, les limites des services écosystémiques.* Versailles: Quae.

———— (2018). *La part sauvage du monde. Penser la nature dans l'Anthropocène.* Paris: Seuil.

Marranca, Bonnie (1996). *Ecologies of Theater: Essays at the Century Turning.* New York: PAJ Publications.

Marson, Anna (ed) (2019). *Urbanistica e pianificazione nella prospettiva territorialista.* Macerata: Quodlibet.

Mauvaise Troupe (2014). *Constellations. Trajectoires révolutionnaires du jeune 21e siècle.* Paris: Éditions de L'éclat.

———— (2017). *Saisons – nouvelles de la zad.* Paris: Éditions de L'éclat.

McCartney, Andra, and David Paquette (2012). "Soundwalking and the Bodily Exploration of Places". *Canadian Journal of Communication* 37(1): 135–45.

Meadows, Donella, Dennis Meadows, Jorgen Randers, and William W. Beherens (1972). *The Limits to Growth.* New York: Universe Books.

Mekdjian, Sarah, and Elise Olmedo (2016). "Médier des récits de vie. Expérimentation de cartographies narratives et sensibles". *M@ppemonde* 118. URL: https://shs.hal.science/halshs-01242536

Merabet, Emma, Anne-Sophie Noel, and Julie Sermon (eds) (2019). *Matières.* Special Issue of *Agôn* 8. URL: https://journals.openedition.org/agon/5801

Meusburger, Peter, Joachim Funke, and Edgar Wunder (eds) (2009). *Milieus of Creativity: An Interdisciplinary Approach to Spatiality of Creativity.* Heidelberg: Springer.

Ministère de la culture (2020). "L'impact de la crise du Covid-19 sur les secteurs culturels". URL: https://www.culture.gouv.fr/Thematiques/Etudes-et-statistiques/Publications/Collections-de-synthese/Culture-chiffres-2007-2022/L-impact-de-la-crise-du-Covid-19-sur-les-secteurs-culturels (accessed 5 September 2020).

Moeglin-Delcroix, Anne (2006). "Au-delà du langage. La poésie selon herman de vries", in *Sur le livre d'artiste. Articles et écrits de circonstance (1981–2005).* Marseille: Le mot et le Areste.

Mondzain, Marie-José (2005). *Image, Icon, Economy: The Byzantine Origins of the Contemporary Imaginary,* trans. Rico Franses. Stanford: Stanford University Press.

Montano, Linda (2011). "Entretien avec Ana Mendieta", in *Ana Mendieta. Blood and Fire,* Abigail Solomon-Godeau, Linda Montano, Nancy Princenthal, and Mary Sabbatino (eds) Paris: Galerie Lelong, 28.

Moreau, Yoann (2017). *Vivre avec les catastrophes.* Paris: Presses Universitaires de France.

Morin, Edgar (1977). *La Méthode 1: La nature de la nature.* Paris: Seuil.

———— ([1982] 1990). *Science avec Conscience.* Paris: Fayard/Seuil.

———— (2004). *La Méthode 6: Éthique.* Paris: Seuil.

—— (2008). *On Complexity*. New York: Hampton Press.

Morizot, Baptiste (2016). *Les Diplomates. Cohabiter avec les loups sur une autre carte du vivant*. Marseille: Editions Wildproject.

—— (2018). *Sur la piste animale*. Arles: Actes Sud.

Morizot, Baptiste, and Estelle Zhong Mengual (2018). *Esthétique de la rencontre*. Paris: Seuil.

Morris, Robert (1966). "Notes on Sculpture. Part II". *Artforum* 5(2): 20–3.

Morton, Timothy (2009). *Ecology without Nature. Rethinking Environmental Aesthetics*. Cambridge, MA and London: Harvard University Press.

—— (2010). *The Ecological Thought*. Cambridge, MA and London: Harvard University Press.

—— (2018). *Hyperobjets. Philosophie et écologie après la fin du monde*. Saint-Étienne: Cité du Design.

—— (2019). *Being Ecological*. Cambridge, MA: MIT Press.

—— (2021). *La Pensée écologique*. Paris: Zulma.

Mottet, Jean (1998). *L'invention de la scène américaine. Cinéma et paysage*. Paris: L'Harmattan.

Moulier-Boutang, Yann (2012). *Cognitive Capitalism*, trans. Ed Emery. Cambridge: Polity Press.

Næss, Arne (1973). "The Shallow and the Deep, Long-Range Ecology Movement. A Summary". *Inquiry* 16: 95–100.

Nancy, Jean-Luc (2021). "From ontology to technology", in *The Fragile Skin of the World*, trans. Cory Stockwell. Cambridge: Polity Press.

Naoufal, Nayla (2018). "Danser avec des vers de farine: transcorporéité et sollicitude". Paper presented at the *Arts, Ecologies et Transitions Conference*, Paris (12 October 2018). URL: http://www-artweb.univ-paris8.fr/?Premieres-Rencontres-Arts

Narraway, Guinevere, and Anat Pick (eds) (2013). *Screening Nature: Cinema Beyond the Human*. New York: Berghahn Books.

Neidich, Warren (2013). "The early and late stages of cognitive capitalism", in *The Psychopathologies of Cognitive Capitalism: Part Two*, Warren Neidich (ed) Berlin: Archive Books, 9–28.

Neuhaus, M. (2019). *Les pianos ne poussent pas sur les arbres*. Dijon: Les Presses du Réel.

Neyrat, Frédéric (2015). "Le cinéma éco-apocalyptique. Anthropocène, cosmophagie, anthropophagie". *Communications* 96: 67–79.

Norman, Katharine (2012). "Listening Together, Making Place". *Organised Sound* 17(3): 257–65.

Nova, Nicolas (ed) (2021). *A Bestiary of the Anthropocene. Hybrid Plants, Animals, Minerals, Fungi, and other Specimens*. Eindhoven: Onomatopee.

Ostrom, Elinor (1990). *Governing the Commons: The Evolution of Institutions for Collective Action*. Cambridge: Cambridge University Press.

Papaeti, Anna (2013). "Music, Torture, Testimony: Reopening the Case of the Greek Military Junta (1967–1974)". *The World of Music* 2(1): 67–90.

Paparrigopoulos, Kostas (2016a). "Décroissance, musique et environnement sonore: relations et interdépendances", in *Art i decreixement/Arte y decrecimiento/Art et décroissance*, Carmen Pardo (ed) Girone: Documenta Universitaria, 141–60.

—— (2016b). "Sons désirables et sons indésirables: Une dichotomie avec plusieurs extensions", in *Musique et écologies du son. Propositions théoriques pour une écoute du monde*, Makis Solomos, Roberto Barbanti, Guillaume Loizillon, Kostas Paparrigopoulos, and Carmen Pardo (eds) Paris: Éditions L'Harmattan.

Paquot, Thierry, and Chris Younès (2009). *Le territoire des philosophes. Lieu et espace dans la pensée du XXe siècle*. Paris: La Découverte.

Paquot, Thierry (ed) (2012). *Espace et lieu dans la pensée occidentale*. Paris: La Découverte.

Pardo, Carmen (1999). "Comment faire tenir un espace? Entretien", in *Pascale Criton. Les univers microtempérés*. Champigny sur Marne: À la ligne, 56–68.

—— (2016a). "Prácticas musicales y ecología mental", in *Art i decreixement/Arte y decrecimiento/Art et décroissance*, Carmen Pardo (ed) Girone: Documenta Universitaria, 175–94.

—— (ed) (2016b). *Art i decreixement/Arte y decrecimiento/Art et décroissance*. Girone: Documenta Universitaria.

—— (2017). "The Emergence of Sound Art: Opening the Cages of Sound". *The Journal of Aesthetics and Art Criticism* 75(1): 35–48.

Parikka, Jussi (2019). "Cartographies of Environmental Arts", in *Posthuman Ecologies: Complexity and Process after Deleuze*, Rosi Braidotti and Simon Bignall (eds) London: Rowman & Littlefield.

Paris-Clavel, Gérard (2018). "Entretien avec Yann Aucompte", in *La diversité éthique des pratiques de graphisme en France au tournant du XXIe siècle*, Yann Aucompte, PhD dissertation. Université Paris 8, France: Annexe 6.6: 33–5.

Passeron, René (2000). *Pour une philosophie de la création*. Paris: Klincksieck.

Pater, Ruben (2021). *Caps Lock*. Amsterdam: Valiz.

Patriarca, Éliane (ed) (2015). *Du souffle dans les mots. Parlement sensible – Trente écrivains s'engagent pour le climat*. Paris: Arthaud.

Patti, Daniela, and Levente Polyak (2015). "From practice to policy: frameworks for temporary use". *Urban Research & Practice* 8(1): 122–34.

Pauly, Daniel (1995). "Anecdotes and the shifting baseline syndrome of fisheries". *Trends in Ecology & Evolution* 10(10): 430.

Payant, René (1981). "Le Choc du présent. Notes sur la performance: l'effet cinéma", in *Performance Text(e)s & Documents*, Chantal Pontbriand (ed) Montréal: Parachute, 127–37.

Pérec, Georges (1989). *L'infra-ordinaire*. Paris: La Librairie du XXIe siècle.

Perrin, Julie (2017). "Traverser la ville ininterrompue: sentir et se figurer à l'aveugle. À propos de *Walk, Hands, Eyes (a city)* de Myriam Lefkowitz". *Ambiances* 3. URL: http://ambiances.revues.org/962

—— (2019). *Questions pour une étude de la chorégraphie située*, HDR dissertation. Université de Lille, France. URL: https://univ-paris8.hal.science/tel-02208299v2

Perrottet, Vincent (2012). "Ouvrez l'œil", in *Affiche-Action: quand la politique s'écrit dans la rue*, Béatrice Fraenkel, Magali Gouiran, Nathalie Jakobowicz, and Valérie Tesnière (eds) Paris: Gallimard, 126–35.

Peschier-Pimont, Laurie, and Lauriane Houbey (2019). *Waving – Tracés Chorégraphiques*. Nantes: Inui collectif.

Petit, Victor (2013). "Le concept de milieu en amont et en aval de Simondon". *Cahiers Simondon* 5: 45–58.

—— (2015). "L'éco-design: design de l'environnement ou design du milieu?". *Sciences du design* 2: 31–9.

Pierrepont, Alexandre (2021). *Chaos, cosmos, musique*. Paris: Éditions MF.

Pignocchi, Alessandro (2019). *La recomposition des mondes*. Paris: Seuil.

Pinson, Jean-Claude (2020). *Pastoral. De la poésie comme écologie.* Ceyzérieu: Champ Vallon.

Plumwood, Val (1993). *Feminism and the Mastery of Nature.* London: Routledge.

Poinsot, Jean-Marc (1986). "In situ, lieux et espaces de la sculpture contemporaine", in *Qu'est-ce que la sculpture moderne?* Margit Rowell (ed) Paris: Centre Georges Pompidou, 322–9.

—— (1989). "L'in situ et la circonstance de sa mise en vue". *Les Cahiers du Musée national d'art moderne* 27: 66–75.

—— (2008). *Quand l'œuvre a lieu. L'art exposé et ses récits autorisés.* Genève: Les Presses du Réel, 80–109.

Poisson, Mathias (2010). "Le réel me perce". *Les Cahiers de Sentier* 3: 20–2.

Poivert, Michel (2018). *La Photographie contemporaine.* Paris: Flammarion.

—— (2022). *Contre-culture dans la photographie.* Paris: Textuel.

Posthumus, Stéphanie (2010). "État des lieux de la pensée écocritique française". *Ecozon@* 1(1): 148–54.

Prampolini, Enrico (1926). "The Magnetic Theatre and the Futuristic Scenic Atmosphere", trans. Rosamond Gilder. *The Little Review*: 101–8.

—— ([1924] 1973). "L'atmosphère scénique futuriste", in *Le Futurisme. Manifestes – Proclamations – Documents*, Giovanni Lista (ed) Lausanne: Editions L'Âge d'Homme, 281–4.

Prévieux, Julien (2017). *Of Balls, Books and Hats.* Performance. France.

Pyle, Robert (2003). "Nature matrix: reconnecting people and nature". *Oryx* 37(2): 206–14.

—— (2016). "L'extinction de l'expérience", trans. Mathias Lefèvre. *Écologie & Politique* 53(2): 185–96.

Qadir, Junaid, Arjuna Sathiaseelan, Liang, Wang, and Jon Crowcroft (2016). "Taming limits with approximate networking", in *LIMITS '16: Proceedings of the Second Workshop on Computing within Limits*, Bonnie Nardi (ed) New York: Association for Computing Machinery, 1–8.

Quoiraud, Christine, and Hamish Fulton (2003). *Walk Dance Art C°.* Trézélan: Filigranes.

Radjou, Navi, and Jaideep Prabhu (2015). *Frugal Innovation: How to do better with less.* India: Hachette.

Raharimanana, Jean-Luc (2008). *Za.* Paris: Philippe Rey.

Ramade, Bénédicte (2022). *Vers un art anthropocène. L'art écologique américain pour prototype.* Dijon: Les Presses du Réel.

Rancière, Jacques (2004). *Aux bords du politique.* Paris: Gallimard Folio Essais.

—— (2009). *The Emancipated Spectator*, trans. Gregory Elliott. London: Verso.

—— (2010). *Dissensus: On Politics and Aesthetics*, trans. Steven Corcoran. London: Continuum.

—— (2011). *Aisthesis. Scenes from the Aesthetic Regime of Art.* London/New York: Verso.

—— (2018). *La méthode de la scène.* Paris: Ligne.

Rasmi, Jacopo (2019). *Écologie des méthodes documentaires à partir d'écritures filmiques et littéraires de l'Italie contemporaine*, PhD dissertation. Université Grenoble Alpes, France.

—— (2021). *Le hors-champ est dedans! Michelangelo Frammartino, écologie, cinéma.* Lille: Presses du Septentrion.

Récopé, Michel, Géraldine Rix-Lièvre, Hélène Fache, and Simon Boyer (2013). "La *sensibilité à*, organisatrice de l'expérience vécue", in *Expérience, activité, apprentissage*, Luc Albarello, Jean-Marie Barbier, Étienne Bourgeois, and Marc Durand (eds) Paris: Presses Universitaires de France, 111–33.

Reyna, Alejandro (2018). "Musique et expériences: éthique du partage des expériences dans *Chantal, ou le portrait d'une villageoise* de Luc Ferrari". *Filigrane. Musique, esthétique, sciences, société* (23). URL: https://revues.mshparisnord.fr:443/filigrane/index.php?id=891

Reynié, Dominique (2012). "Auguste Comte l'opinion publique par l'affiche", in *Affiche-Action: quand la politique s'écrit dans la rue*, Béatrice Fraenkel, Magali Gouiran, Nathalie Jakobowicz, and Valérie Tesnière (eds) Paris: Gallimard, 38–45.

Richards, John (2017). "DIY and maker communities in electronic music", in *The Cambridge Companion to Electronic Music*, Nick Collins and Julio d'Escriván (eds) Cambridge: Cambridge University Press, 238–57.

Ricœur, Paul (2001). *Memory, History, Forgetting*, trans. Kathleen Blamey and David Pellauer. Chicago: Chicago University Press.

Rock, Michael ([1996] 2013). "Designer as author", in *Multiple Signatures*. New York: Rizzoli, 45–56.

Roquet, Christine (2019). *Vu du geste, Interpréter le mouvement dansé*. Paris: CND.

Rosa, Hartmut (2019). *Resonance: A Sociology of Our Relationship to the World*, trans. James Wagner. Cambridge: Polity Press.

Rose, Arthur R. (1969). "Four Interviews with Barry, Huebler, Kosuth, Weiner". *Arts Magazine* 43(4): 22–3.

Roulier, Frédéric (1999). "Pour une géographie des milieux sonores". *Cybergeo: European Journal of Geography* 71. URL: http://cybergeo.revues.org/5034 (accessed 1 February 2017).

Roustang, Guy (2006). "Décroissance", in *Dictionnaire de l'autre économie*, Jean-Louis Laville and Antonio David Cattani (eds) Paris: Gallimard, 144–51.

Rueckert, William (1978). "Literature and Ecology: An Experiment in Ecocriticism". *Iowa Review* 9(1): 71–86.

Rumpala, Yannick (2019). *Hors des décombres du monde: écologie, science-fiction et éthique du futur*. Ceyzérieu: Champ Vallon.

Russell, Ben (ed) (2007). *Black and White Trypps Number Three*. URL: https://lightcone.org/fr/film-4963-black-and-white-trypps-number-three

Rust, Stephen, Salma Monani, and Sean Cubitt (eds) (2013). *Ecocinema Theory and Practice*. New York: Routledge.

Sadin, Eric (2021). *L'Intelligence artificielle ou l'enjeu du siècle. Anatomie d'un antihumanisme radical*. Paris: L'Echappée.

Sainte-Beuve, Charles-Augustin (1862). "Du Génie critique de Bayle", in *Portraits littéraires*, vol. 1. Paris: Garnier Frères, 364–88.

Saladin, Matthieu (2020). "The Inaudible as an Effect: Tactics of Sound Erasure in Max Neuhaus", in *Sound Unheard*, Anne Zeitz (ed) Special Issue of *Kunsttexte.de* 1. URL: https://journals.ub.uni-heidelberg.de/index.php/kunsttexte/issue/view/6090

Schafer, Raymond Murray (1977). *The Soundscape, The Tuning of the World*. Vermont: Destiny Books.

——— (ed) (2009). *Listen*. URL: http://www.nfb.ca/film/listen

Schoentjes, Pierre (2020). *Littérature et écologie. Le mur des abeilles*. Paris: José Corti.

Schulz, Bernd (1999). "Einleitung", in Robin Minard, *Silent Music*, Bernd Schulz (ed) Heidelberg: Kehrer Verlag.

Scott, Allen J. (2010). "Creative cities: the role of culture". *Revue d'Economie Politique* 120: 181–204.

Sealy, Mark (2019). *Decolonising the Camera: Photography in Racial Time*. London: Lawrence & Wishart Ltd.

Semal, Luc (2019). *Face à l'effondrement. Militer à l'ombre des catastrophes*. Paris: PUF.

Sennet, Richard (1998). *The Corrosion of Character, The Personal Consequences of Work in the New Capitalism*. New York: W. W. Norton.

Sermon, Julie (2018). "Les imaginaires écologiques de la scène actuelle. Récits, formes, affects". *Théâtre/Public* 229: 4–11.

——— (2021). *Morts ou vifs. Contribution à une écologie pratique, théorique et sensible des arts vivants*. Paris: B42.

Serra, Richard (1980). "Richard Serra's urban sculpture: An interview Richard Serra & Douglas Crimp", in *Richard Serra: Interviews, Etc. 1970–1980*, Clara Weyergraf-Serra (ed) New York: The Hudson River Museum, 164–87.

Serre, Joséphine (2019). *Data Mossoul*. Montreuil: Éditions Théâtrales.

Serres, Michel (2009). *Écrivains, savants et philosophes font le tour du monde*. Paris: Le Pommier.

Servigne, Pablo, and Raphaël Stevens (2015). *Comment tout peut s'effondrer. Petit manuel de collapsologie à l'usage des générations présentes*. Paris: Seuil.

Servigne, Pablo, Raphaël Stevens, and Gauthier Chapelle (2018). *Une autre fin du monde est possible. Vivre l'effondrement (et pas seulement y survivre)*. Paris: Seuil.

Shepard, Paul (2013). *Nous n'avons qu'une seule Terre*. Paris: J. Corti.

Shepard (ed) (1996). *The Only World We've Got*. San Francisco: Sierra Club Books.

Shore, Robert (2014). *Post-Photography, The Artist with a camera*. London: Laurence King Publishing.

Sicard, Monique (1998). "L'image-calcul, dans Josette Sultan, Jean-Christophe Vilatte, Ce corps incertain de l'image". *Art/Technologies, Champs Visuels* 10.

Sierek, Karl (2013). "Animisme de l'image. Pour une histoire de la théorie d'un concept mouvant". *Intermédialités/Intermediality* 22.

Simon, Roger I. (1987). "Empowerment as a Pedagogy of Possibility". *Language Arts* 64(4): 370–82.

Simondon, Gilbert (2005). *L'individuation à la lumière des notions de forme et d'information*. Paris: Million.

——— ([1958] 2012). *Du mode d'existence des objets techniques*. Paris: Aubier.

——— (2015). *L'individuation à la lumière des notions de forme et d'information*. Grenoble: Million.

——— (2020). *Individuation in Light of Notions of Form and Information*, trans. Taylor Adkins. Minneapolis and London: University of Minnesota Press.

Solomos, Makis (2005). "Cellular automata in Xenakis's music. Theory and practice", in *Proceedings of the International Symposium Iannis Xenakis*, Anastasia Georgaki and Makis Solomos (eds) Athens: Université d'Athènes.

——— (2016). "Musique et décroissance. Une première approche", in *Art i decreixement/Arte y decrecimiento/Art et décroissance*, Carmen Pardo (ed) Girone: Documenta Universitaria, 97–112.

———— (2018a). "From sound to sound space, sound environment, soundscape, sound milieu or ambiance", in *Soundings and Soundscapes*, Sarah Kay and François Noudelmann (eds) Special Issue of *Paragraph. A Journal of Modern Critical Theory* 41(1): 95–109.

———— (2018b). "L'écoute musicale comme construction du commun". *Circuit. Musiques Contemporaines* 8(3): 54–64.

Solomos, Makis, Roberto Barbanti, Guillaume Loizillon, Kostas Paparrigopoulos, and Carmen Pardo (eds) (2016). *Musique et écologies du son. Propositions théoriques pour une écoute du monde*. Paris: L'Harmattan.

———— (2020). *Form Music to Sound. The Emergence of Sound in 20th- and 21st-Century Music*. London: Routledge.

———— (2023). *Exploring the Ecologies of Music and Sound. Environmental, Mental and Social Ecologies in Music, Sound Art and Artivisms*. London: Routledge.

Sontag, Susan (1965). "The Imagination of Disaster". *Commentary* 40(4): 42–8.

Sorin, Cécile (2017a). "L'écologie linguistique pasolinienne à l'œuvre dans le cinéma plurilingue français contemporain", in *Pier Paolo Pasolini: entre art et philosophie*, Véronique Le Ru and Fabrice Bourlez (eds) Reims: ÉPURE, 63.

———— (2017b). *Pasolini, pastiche et mélange*. Saint-Denis: Presses universitaires de Vincennes.

Sorin, Cécile, and Emna Mrabet (2021). "L'Hétérotope. Du 'film guérilla'". *Création Collective au Cinéma* 5: 243–58.

Spaid, Sue (2002). *Ecoventions: Current Art to Transform Ecologies*. Cincinnati: Greenmuseum.org and The Contemporary Art Center: Ecoartspace.

Stachelhaus, Heiner (1994). *Joseph Beuys. Une biographie*. Paris: Abbeville.

Stein, Gertrude (1988). *Lectures in America*. London: Virago.

———— ([1935] 2011). *Lectures en Amérique*, trans. Claude Grimal. Paris: Christian Bourgois.

Stengers, Isabelle (2015). *In Catastrophic Times. Resisting the Coming Barbarism*, trans. Andrew Goffey. London: Open Humanities Press.

———— (2019). *Résister au désastre*. Marseille: Editions Wildproject.

Stengers, Isabelle, and Serge Gutwirth (2018). *Pourquoi ce qui se passe à Notre-Dame-des-Landes nous importe-t-il?*. URL: https://groupeconstructiviste.wordpress.com/2018/05/11/pourquoi-ce-qui-se-passe-a-notre-dame-des-landes-nous-importe-t-il/#_ftn3

Stern, Daniel ([1985] 1998). *The Interpersonal World of the Infant: A View from Psychoanalysis and Developmental Psychology*. London: Karnac Books.

Steyerl, Hito (2012). *The Wretched of the Screen*. Berlin: Sternberg Press.

Stiegler, Bernard (1994). *La Technique et le Temps 1*. Paris: Galilée.

———— (1996). *La Technique et le Temps 2. La désorientation*. Paris: Galilée.

Suberchicot, Alain (2012). *Littérature et environnement: Pour une écocritique comparée*. Paris: Honoré Champion.

Szendy, Peter (2017). *Apocalypse-Cinema: 2012 and Other Ends of the World*, trans. Will Bishop. New York: Fordham University Press.

———— (2019). *The Supermarket of the Visible: Toward a General Economy of Images*. New York, NY: Fordham University Press.

Tassin, Jacques, and Christian A. Kull (2012). "Pour une autre représentation métaphorique des invasions biologiques". *Natures Sciences Sociétés* 20(4): 404–14.

Taylor, Diana (2003). *The Archive and the Repertoire: Performing Cultural Memory in the Americas*. Durham: Duke University Press.

Tesnière, Valérie (2012). "'De l'affichage politique' conseils pratiques de l'abbé Fourié", in *Affiche-Action: quand la politique s'écrit dans la rue*, Béatrice Fraenkel, Magali Gouiran, Nathalie Jakobowicz, and Valérie Tesnière (eds) Paris: Gallimard, 100–3.

Thiong'o, Ngũgĩ wa (1986). *Decolonising the Mind: the Politics of Language in African Literature*. London: J. Currey.

Tiqqun (2009). *Contributions à la guerre en cours*. Paris: La Fabrique.

Todorov, Tzvetan (1982). "Présentation", in *Littérature et réalité*, Gérard Genette and Tzvetan Todorov (eds) Paris: Seuil, 7–10.

Tonkiss, Fran (2013). "Austerity urbanism and the makeshift city". *City* 17(3): 312–24.

Toop, David (2000). *Sonic Boom: The Art of Sound*. London: Hayward Gallery.

Touam Bona, Denetem (2021). *Sagesse des lianes, Cosmopoétique du refuge 1*. Paris: Post Éditions.

——— (2022). *Emprunter tous les rôles possibles, échapper à la prise*. Marseille: Mille Cosmos.

Truax, Barry (2008). "Soundscape Composition as Global Music: Electroacoustic Music as Soundscape". *Organised Sound* 13(2): 103–9.

——— (online). "Soundscape Composition". URL: https://www.sfu.ca/~truax/scomp. html (accessed 11 July 2023).

Tsing, Anna (2015). *The Mushroom at the End of the World. On the Possibility of Life in Capitalist Ruins*. Princeton, NJ: Princeton University Press.

Turri, Eugenio (1979). *Semiologia del paesaggio italiano*. Milan: Longanesi.

Ultra-red (2014). *URXX Nos. 1–9. Nine Workbooks 2010–2014*. Köln: König.

UNESCO (2001). "First Proclamation of Masterpieces of the Oral and Intangible Heritage of Humanity". URL: https://unesdoc.unesco.org/ark:/48223/pf0000124206 (accessed 24 July 2023).

United Nations (2018). "Revision of World Urbanization Prospects". URL: https:// population.un.org/wup/

Valentine, Vincent (2003). "Éducation musicale et ERE: Éléments d'un cadre axiologique". *Vertigo* 2(4). URL: https://doi.org/10.4000/vertigo.4482

Dooren, Thom van, Eben Kirksey, and Ursula Münster (2016). "Multispecies Studies. Cultivating Arts of Attentiveness". *Environmental Humanities* 8(1): 1–23.

van Eck, Cathy (2016). *Between Air and Electricity Microphones and Loudspeakers as Musical Instruments*. London: Bloomsbury.

van Eikels, Kai (2018). "Performance collective, performance de la collectivité: qu'est-ce que l'être-ensemble?", in *Scènes en partage*, Eliane Beaufils and Alix de Morant (eds) Montpellier: Editions Deuxième Époque, 35–47.

van Gennep, Arnold ([1909] 1981). *Les Rites de passage*. Paris: Picard.

Vieillescazes, Nicolas (2019). "Qu'est-ce qu'un intellectuel d'ambiance?". *lundimatin* 189. URL: https://lundi.am/Qu-est-ce-qu-un-intellectuel-d-ambiance-Nicolas-Vieillescazes (accessed 22 January 2023).

Vignola, Gabriel (2017). "Écocritique, écosémiotique et représentation du monde en littérature". *Cygne noir* 5: 11–36. URL: https://id.erudit.org/iderudit/1089937ar (accessed 6 July 2021).

Villani, Tiziana (2019). *Corps mutants. Technologies de la sélection de l'humain et du vivant*. Paris: Eterotopia France.

Voegelin, Salomé (2010). *Listening to Noise and Silence. Towards a Philosophy of Sound Art*. New York: Continuum.

Volcler, Juliette (2011). *Le son comme arme. Les usages policiers et militaires du son*. Paris: La Découverte.

Volvey, Anne (2007). "Land Arts. Les fabriques spatiales de l'art contemporain". *T.I.G.R.* 129–30: 3–25.

Uexküll, Jakob Von (1957). "A stroll through the worlds of animals and men", in *Instinctive Behavior. The Development of Modern Concept*, Claire H. Schiller (ed) New York: International University Press, 5–80.

——— (2010). *A Foray into the Worlds of Animals and Humans*, trans. Joseph D. O'Neil. Minneapolis and London: University of Minnesota Press.

Vries, Herman de ([1972] 2002). *My Poetry Is the World*. Eschenau and Paris: Eschenau Editions/Lydia Megert.

——— (2001). *herman de vries les choses mêmes*. Lyon: Réunion des Musées Nationaux/ Musée départemental de Digne.

Wallin, Nils (1991). *Biomusicology: Neurophysiological, Neuropsychological and Evolutionary Perspectives on the Origins and Purposes of Music*. Stuyvesant, NY: Pendragon Press.

Warren, Karen J. (2009). "The Power and the Promise of Ecological Feminism". *Multitudes* 36(1): 170–6.

Weibel, Peter (1994). "Kontextkunst – Zur sozialen Konstruktion von Kunst", in *Kontext Kunst. Kunst der 90er Jahre*, Peter Weibel (ed) Köln: DuMont Buchverlag. [English translation by Frances Loeffler, in Claire Doherty (ed) (2009). *Situation. Documents of Contemporary Art*. Cambridge, MA and London: MIT Press/ Whitechapel Gallery, 46–52.]

Weil, Eva (2000). "Silence et latence". *Revue française de la psychanalyse* 64(1): 169–82.

Weinberger, Lois (1997). *Notes from the Hortus*. Ostfildern: Cantz Verlag.

——— (2009). *The Mobile Garden*. Bologna: Damiani.

Westerkamp, Hildegard (1988). *Listening and Soundmaking: A Study of Music-As-Environment*, MA dissertation. Simon Fraser University, Canada.

——— (2002). "Linking Soundscape Composition and Acoustic Ecology". *Organised Sound* 7(1): 51–6.

——— (2006). "Soundwalking as ecological practice", in *The West Meets the East in Acoustic Ecology*, Keiko Torigoe, Tadahiko Imada, and Kozo Hiramatsu (eds) Hirosaki: Hirosaki University, 84–9.

Wiame, Aline (2016). *Scènes de la défiguration. Quatre propositions entre théâtre et philosophie*. Dijon: Les Presses du Réel.

Wilson, Edward Osborne (ed) (1988). *BioDiversity*. Washington, DC: National Academy Press. URL: https://nap.nationalacademies.org/catalog/989/biodiversity

Wittig, Monique (1969). *Les Guerrillères*. Paris: Éditions de Minuit.

Wood, Catherine (2018). *Performance in Contemporary Art*. London: Tate Publishing.

World Economic Forum (2020). "This is how COVID-19 is affecting the music industry". URL: https://www.weforum.org/agenda/2020/05/this-is-how-covid-19-is-affecting-the-music-industry/ (accessed 5 September 2020).

Woynarski, Lisa (2020). *Ecodramaturgies: theatre, performance and climate change*. New York: Palgrave MacMillan.

Wright, Erik Olin (2010). *Envisioning Real Utopias*. London: Verso.

Yusoff, Kathryn (2018). *A Billion Black Anthropocenes or None. Forerunners*. Minneapolis: University of Minnesota Press.

Zabunyan, Dork (2019). "À quoi servent les images de la catastrophe écologique?". *Analyse Opinion Critique*. URL: https://aoc.media/opinion/2019/10/30/a-quoi-servent-les-images-de-la-catastrophe-ecologique/

Zalasiewicz, Jan, et al. (2011). "Stratigraphy of the Anthropocene". *Philosophical Transactions: Mathematical, Physical and Engineering Sciences* 369(1938): 1036–55.

Zhong Mengual, Estelle (2019). *L'art en commun. Réinventer les formes du collectif en contexte démocratique*. Dijon: Les Presses du Réel.

Zittel, Andrea, and Emily Mast (2016). "Les pouvoirs secrets de super-héros", in *Savoir utopique, pédagogie radicale et artist-run community art space en Californie du sud*, Géraldine Gourbe (ed) Rennes and Annecy: Shelter Press/ESAAA Editions.

Zuboff, Shoshana (2019). *The Age of Surveillance Capitalism. The Fight for a Human Future at the New Frontier of Power*. New York: Public Affairs.

Web Links

Bourges, Gaëlle (2018). "Ce que tu vois", *Association OS*. URL: https://www.gaellebourges.com/spectacle/ce-que-tu-vois/

Chariatte, Eve. URL: https://evechariatte.ch

CNRS (2018). "Where have all the farmland birds gone?". URL: https://news.cnrs.fr/articles/where-have-all-the-farmland-birds-gone

Crisp, Rosalind "DIRt". URL: https://www.omeodance.com/dirt

——— "L'entretien par C. Astier". URL: https://www.maculture.fr/entretiens/dirt-rosalind-crisp

Devigon, Armelle. Cie LLE. URL: http://www.compagnielle.fr/

Ferrara, Patricia, and Unber Humber Group. "Gestes de terres". URL: https://www.patriciaferrara.org/tag/gestes-de-terre/

Guérédrat, Anabel. "Decolonial and Ecofeminist Ritual with Sargassa". URL: https://artincidence.fr/

ILAND. URL: http://www.ilandart.org/

INUI A choreographer duet working with landscape. URL: https://www.meteores.org/inui-bio

IPBES (Intergovernmental Science-Policy Platform on Biodiversity and Ecosystem Services). URL: https://ipbes.net

IUCN (2020). URL: https://www.iucnredlist.org/

Kerminy, An art and agriculture place. URL: https://kerminy.org

Lang, Prue. URL: https://www.pruelang.com/2013/green-guidelines-compagnie-prue-lang/

Myrtil, Marlène, and The Kaméléonite group. "Chroniques agricoles". URL: https://www.kameleonite.net/chroniques-agricoles

Olsen, Andrea. URL: https://andrea-olsen.com or http://www.body-earth.org

OXFAM (2020). "Davos 2020: Nouveau rapport d'Oxfam sur les inégalites mondiales". URL: https://www.oxfamfrance.org/communiques-de-presse/davos-2020-nouveau-rapport-doxfam-sur-les-inegalites-mondiales/

Pagès, Laurence (2017). "Pour qui tu te prends?". URL: https://laurencepages.wordpress.com/a-propos/

Recoil Performance Group (2018). "MASS – Bloom Explorations". URL: https://recoil-performance.org/productions/mass-bloom-explorations/

Renarhd, Camille. URL: https://camillerenarhd.com/

United Nations (1992). "Convention on Biodiversity". URL: https://www.cbd.int/doc/legal/cbd-en.pdf

Vigie Nature. "Produire des indicateurs à patir des indices des espèces". URL: https://www.vigienature.fr/page/produire-des-indicateurs-partir-des-indices-des-especes-habitat

World Wildlife Fund (2018). "Living Planet Report 2018". URL: https://www.worldwildlife.org/pages/living-planet-report-2018

INDEX

Pages followed by "n" refer to notes.

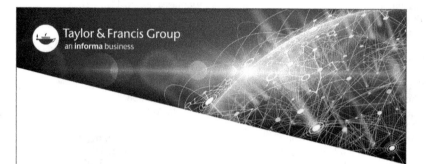